LANGUAGE AND CROSS-CULTURAL COMMUNICATION IN TRAVEL AND TOURISM

Strategic Adaptations

LANGUAGE AND CROSS-CULTURAL COMMUNICATION IN TRAVEL AND TOURISM

Strategic Adaptations

Edited by
Soumya Sankar Ghosh, PhD
Debanjali Roy
Tanmoy Putatunda, PhD
Nilanjan Ray, PhD

First edition published 2025

Apple Academic Press Inc.
1265 Goldenrod Circle, NE,
Palm Bay, FL 32905 USA

760 Laurentian Drive, Unit 19,
Burlington, ON L7N 0A4, CANADA

CRC Press
2385 NW Executive Center Drive,
Suite 320, Boca Raton FL 33431

4 Park Square, Milton Park,
Abingdon, Oxon, OX14 4RN UK

© 2025 by Apple Academic Press, Inc.

Apple Academic Press exclusively co-publishes with CRC Press, an imprint of Taylor & Francis Group, LLC

Reasonable efforts have been made to publish reliable data and information, but the authors, editors, and publisher cannot assume responsibility for the validity of all materials or the consequences of their use. The authors are solely responsible for all the chapter content, figures, tables, data etc. provided by them. The authors, editors, and publishers have attempted to trace the copyright holders of all material reproduced in this publication and apologize to copyright holders if permission to publish in this form has not been obtained. If any copyright material has not been acknowledged, please write and let us know so we may rectify in any future reprint.

Except as permitted under U.S. Copyright Law, no part of this book may be reprinted, reproduced, transmitted, or utilized in any form by any electronic, mechanical, or other means, now known or hereafter invented, including photocopying, microfilming, and recording, or in any information storage or retrieval system, without written permission from the publishers.

For permission to photocopy or use material electronically from this work, access www.copyright.com or contact the Copyright Clearance Center, Inc. (CCC), 222 Rosewood Drive, Danvers, MA 01923, 978-750-8400. For works that are not available on CCC please contact mpkbookspermissions@tandf.co.uk

Trademark notice: Product or corporate names may be trademarks or registered trademarks and are used only for identification and explanation without intent to infringe.

Library and Archives Canada Cataloguing in Publication

Title: Language and cross-cultural communication in travel and tourism : strategic adaptations / edited by
    Soumya Sankar Ghosh, PhD, Debanjali Roy, Tanmoy Putatunda, PhD, Nilanjan Ray, PhD.
Names: Ghosh, Soumya Sankar, editor. | Roy, Debanjali, editor. | Putatunda, Tanmoy, editor. | Ray, Nilanjan, 1984- editor.
Description: First edition. | Includes bibliographical references and index.
Identifiers: Canadiana (print) 2024040761X | Canadiana (ebook) 20240407652 | ISBN 9781774915707 (hardcover) |
    ISBN 9781774915783 (softcover) | ISBN 9781003477143 (ebook)
Subjects: LCSH: Tourism—Social aspects. | LCSH: Communication in tourism. | LCSH: Intercultural communication.
    | LCSH: Language and culture.
Classification: LCC G155.A1 L2875 2025 | DDC 338.4/791—dc23

Library of Congress Cataloging-in-Publication Data

Names: Ghosh, Soumya Sankar, editor. | Roy, Debanjali, editor. | Putatunda, Tanmoy, editor. | Ray, Nilanjan, editor.
Title: Language and cross-cultural communication in travel and tourism : strategic adaptations / Edited by
    Soumya Sankar Ghosh, PhD, Debanjali Roy, Tanmoy Putatunda, PhD, Nilanjan Ray, PhD.
Description: First edition. | Palm Bay, FL : Apple Academic Press, 2025. | Includes bibliographical references and index. |
    Summary: "This new volume attempts to illustrate how one of the rapidly evolving industries in the world-travel and tourism-has transcended its immediate economic concerns and has become a major signifier for cultural patterns and cross-cultural communications. It discusses the theoretical and scholarly undertakings of tourism that have steadily grown and how the function of language has become the subject of scrutiny in the context of intellectual deliberation vis-à-vis travel and tourism. Grouped in three sections-Cross-cultural Communication in Travel and Tourism, Narrative of Place and Space, and Travel and Tourism Industry, this volume cross-examines culture and intercultural facets of tourism interactions and focuses on culture as a process or phenomenon engaged in or enacted on by individuals. Drawing on discourse analytics and ethnographic approaches, this volume brings together perspectives from the lived experiences of residents, hosts, and ethnographers to explore the extent to which linguistic and cultural differences are identified, constructed, negotiated, and maintained in tourism encounters. The chapters in this book incorporate global case studies and demonstrate the application of theoretical discourses rooted in various disciplines such as marketing, management processes, literature, media studies, and linguistics. The volume draws on insights from those working across a range of geographic contexts and explores the interplay of these issues in various languages and how ethnic language can be used in tourism interactions. The book also addresses the repercussions of the Covid-19 pandemic on the tourism industry and the simultaneous rise in the influence of social media in the cultural semioscape of tourism"-- Provided by publisher.
Identifiers: LCCN 2024025149 (print) | LCCN 2024025150 (ebook) | ISBN 9781774915707 (hardback) |
    ISBN 9781774915783 (paperback) | ISBN 9781003477143 (ebook)
Subjects: LCSH: Tourism--Social aspects. | Communication in tourism. | Intercultural communication. | Language and culture.
Classification: LCC G155.A1 L2875 2025 (print) | LCC G155.A1 (ebook) | DDC 302.2--dc23/eng/20240712
LC record available at https://lccn.loc.gov/2024025149
LC ebook record available at https://lccn.loc.gov/2024025150

ISBN: 978-1-77491-570-7 (hbk)
ISBN: 978-1-77491-578-3 (pbk)
ISBN: 978-1-00347-714-3 (ebk)

About the Editors

Soumya Sankar Ghosh, PhD
Senior Assistant Professor (I), School of Advanced Sciences and Languages, VIT Bhopal University, Madhya Pradesh, India

Soumya Sankar Ghosh, PhD, is Senior Assistant Professor (I) School of Advanced Sciences and Languages, VIT Bhopal University, Madhya Pradesh, India. His teaching experience is in applied linguistics, core linguistics, and communicative and business English. Dr. Ghosh has worked on conversation model building and dialogue processing in the field of machine learning. He is interested in both natural language understanding and natural language generation (NLG) to extract meaning from spoken or written discourse and in feeding NLG systems so that machines can deliver the desired output. He has publications in University Grants Commission CARE-listed, ACL, SCOPUS, CPCI (Web of Science) indexed journals and proceedings. He is a National Advisory Committee Member of the International Conference on Rhythm in Speech and Music and a general member of the Linguistic Society of India. Dr. Ghosh holds a doctorate from the School of Languages and Linguistics at Jadavpur University Kolkata, India. As a research fellow, he has worked at Potsdam University, Germany. He has done his first master's degree in Linguistics at the University of Calcutta, India, and his second master's degree in English Literature at Indira Gandhi National Open University (IGNOU), India. He has earned an International Advanced Diploma Certificate in TESOL/TEFL (Teaching English to Speakers of Other Languages/ Teaching English as a Foreign Language) with a specialization in Business English, from the Asian College of Teachers.

vi

Debanjali Roy

Assistant Professor (I), School of Language and Literature,
KIIT (Deemed to be) University, Odisha, India

Debanjali Roy is working as an Assistant Professor (I), School of Language and Literature, KIIT (Deemed to be) University, Odisha, India. She is pursuing a PhD in the Department of English, University of Calcutta, India. She has a cumulative teaching experience (postgraduate and undergraduate levels) of seven years. A recipient of a T.S. Sterling Scholarship for academic excellence from Presidency College (presently University), Kolkata, she graduated first class in both undergraduate and postgraduate courses of studies in English Literature. Ms. Roy has contributed to ELT and Gender Studies in the Indian context through her field-based research and articles in various peer-reviewed prestigious international journals. Her academic contributions include publications in journals listed in University Grants Commission CARE and indexed by Scopus and Web of Science. She has received professional recognition from the Board of Practical Training, Eastern Region, Ministry of Education, India, and scholarships from international organizations as the Teacher Training Fellowship to the University of Oregon American English Institute (UOAEI), funded by the US Department of State and Office of English Language Programs in 2014 and an Erasmus Mundus Teacher Exchange Fellowship for the year 2020 funded by the European Union. Ms. Roy presently heads the Women's Cell in the School of Language, KIIT Deemed to be University.

Tanmoy Putatunda, PhD

Assistant Professor (I), School of Language and Literature,
KIIT (Deemed to be) University, Odisha, India

Tanmoy Putatunda, PhD is an Assistant Professor (I), School of Language and Literature, KIIT (Deemed to be) University, Odisha, India. He is currently pursuing a PhD from the Department of English, Visva-Bharati, Santiniketan, India. He has a cumulative teaching experience of six years (postgraduate and undergraduate). Prior to joining KIIT University, he served as the Head of the Department of English Language and Literature, Adamas University, Kolkata, India. He completed his bachelor's and master's degrees in English Literature with first class from Visva-Bharati,

Santiniketan. He was awarded a University Grants Commission Junior Research Fellowship (JRF) in 2013. He was engaged in a research project on the "Impact of Social Media on the Lives of the Women of the Kantha Embroidery Industry," undertaken by IIM Kozhikode, India, in 2015. Mr. Putatunda has contributed significantly to the field of urban studies and literature by his lectures and publications in University Grants Commission CARE-listed and SCOPUS-indexed journals and conference proceedings.

Nilanjan Ray, PhD
Associate Professor, Department of Management Studies,
JIS University, West Bengal, India

Nilanjan Ray, PhD, is from Kolkata, India, and presently associated as an Associate Professor in the Department of Management Studies, JIS University, West Bengal, India. Prior to joining JIS University, Dr. Ray was at the Institute of Leadership Entrepreneurship and Development as Associate Professor and Head of Department with additional responsibility as Director IQAC. Prior to that he was also associated at Adamas University as Associate Professor of Marketing Management and Centre Coordinator for Research in Business Analytics at Adamas University in the Department of Management, School of Business and Economics, West Bengal, India. Dr. Ray has obtained a certified Accredited Management Teacher Award from the All India Management Association, New Delhi, India. He has earned the degrees of PhD (Mktg); MCom (Mktg); MBA (Mktg); and STC FMRM (IIT-KGP). He has more than 12 years of teaching and six years of research experience. He has supervised and awarded two doctoral scholars and guided around 56 postgraduate students projects also. Dr. Ray has contributed over 90 research papers in reputed national and international referred, peer-reviewed journals and proceedings and 13 edited research handbooks from Springer, IGI-Global USA, and Apple Academic Publisher CRC Press (A Taylor & Francis Group), USA. He has obtained one patent from Germany and two copyrights from India. He has also associated himself as a reviewer of Tourism Management, Journal of Service Marketing, Journal of Business and Economics, and Research Journal of Business and Management Accounting and as an editorial board member of several referred journals. Dr. Ray has organized several faculty development programs, national and international conferences, and management doctoral colloquiums. Dr. Ray is a life-member of the International Business Studies Academia, fellow member of the Institute of Research Engineers and Doctors Universal Association of Arts and Management Professionals (UAAMP) New York, USA, and Calcutta Management Association (CMA).

Contents

Contributors .... *xi*

Abbreviations .... *xiii*

Acknowledgments .... *xv*

Foreword .... *xvii*

Preface .... *xxi*

Introduction .... *xxiii*

1. **Passenger Identity in Public Transport Awareness Campaign Posters: Contrastive Study of Communication Styles in Japan and France** 1
   Yui Kurihara, Jungah Choi, and Yoshinori Nishijima

2. **How Philosophy of Tourism Tussles with Cultural Diversity and Cultural Tolerance** 33
   Sooraj Kumar Maurya

3. **Intercultural Communication in Tourism During the COVID-19 Pandemic: Analyzing Audience Appeal of Indian Tourism Campaign Through Social Media** 49
   Ruma Saha and Lakhan Raghuvanshi

4. **Deconstructing the Notion of Sacred and Profane from the Viewpoint of Theme-Based Durga Puja in Kolkata in the Age of COVID-19: A Sociological Study** 69
   Soumya Narayan Datta

5. **COVID-19 Pandemic and the End of Overtourism: A Perspective** 91
   Rajdeep Deb and Pankaj Kumar

6. **Transfiguring the Troubled Past Through Narratives** 103
   Duygu Onay Çöker

7. **Literary Ethnography and Travel Aesthetics: Amitav Ghosh's The Hungry Tide and Jungle Nama: A Story of the Sundarbans** 117
   Nobonita Rakshit and Rashmi Gaur

8. Journeying Through the Lesser-Known Indian Spaces:
A Reading of Bishwanath Ghosh's Chai, Chai....................................131
Pamela Pati ànd I Watitula Longkumer

9. Imagined Communities: The Development of the Early
Tourism Industry in Alaska and the Marketing of the
Indigenous Experience...149
Vera Parham and Jennifer Williams

10. Multimodal Approach to Tourism Advertising Discourses.................173
Bui Thi Kim Loan

11. The Saga of Kochi: Cultural and Heritage Tourism Overview..........191
Navneet Munoth, Linson Thomas, and Shubham Gehlot

12. Rural Tourism in India: Constraints and Opportunities....................223
C. Magesh Kumar, K. Sujatha, and K. Rajesh Kumar

13. The Effect of Social Media Sites Promoting Tourism Industry:
An Indian Perspective..235
Vinod Bhatt and Ajay Verma

Index..253

Contributors

Vinod Bhatt
School of Applied Sciences and Languages, VIT Bhopal University, Bhopal, India

Jungah Choi
Institute of Human and Social Sciences, Kanazawa University, Kanazawa, Japan

Duygu Onay Çöker
Department of Visual Communication Design, Faculty of Architecture and Design, TED University, Turkey

Soumya Narayan Datta
Department of Sociology, Bijoy Krishna Girls' College, Howrah, India
Department of Sociology, Adamas University, Kolkata, India

Rajdeep Deb
Department of Tourism & Hospitality Management, Mizoram University, Mizoram, India

Rashmi Gaur
Department of Humanities and Social Sciences, IIT Roorkee, Roorkee, India

Shubham Gehlot
Data Analyst and Urban Planner, Indore, India

C. Magesh Kumar
Department of Business Administration, Annamalai University, Chidambaram, Tamil Nadu, India

Pankaj Kumar
Department of Tourism & Hospitality Management, Mizoram University, Mizoram, India

K. Rajesh Kumar
Department of Business Administration, Annamalai University, Chidambaram, Tamil Nadu, India

Yui Kurihara
Graduate School of Humanities, Osaka University, Minoh, Japan

Bui Thi Kim Loan
Faculty of Foreign Languages, Van Lang University, Ho Chi Minh City, Vietnam

I Watitula Longkumer
Discipline of English, HSS Institute of Infrastructure, Technology, Research and Management (IITRAM), Ahmedabad, India

Sooraj Kumar Maurya
Zakir Husain Delhi College (Evening), University of Delhi, New Delhi, India

Navneet Munoth
Department of Architecture and Planning, Maulana Azad National Institute of Technology, Bhopal, India

Yoshinori Nishijima
Institute of Human and Social Sciences, Kanazawa University, Kanazawa, Japan

Vera Parham
Department of History, American Public University, Charles Town, West Virginia, USA

Pamela Pati
Discipline of English, HSS Institute of Infrastructure, Technology, Research and Management (IITRAM), Ahmedabad, India

Lakhan Raghuvanshi
Department of Journalism & Mass Communication, Devi Ahilya Vishwavidyalaya, Indore, India

Nobonita Rakshit
Department of Humanities and Social Sciences, IIT Roorkee, Roorkee, India

Ruma Saha
Department of Journalism & Mass Communication, Manipal University Jaipur, Rajasthan, India

K. Sujatha
Department of Business Administration, Annamalai University, Chidambaram, Tamil Nadu, India

Linson Thomas
Urban Planner, Kannur, India

Ajay Verma
School of Applied Sciences and Languages, VIT Bhopal University, Bhopal, Madhya Pradesh, India

Jennifer Williams
Department of History, American Public University, Charles Town, WV, USA

Abbreviations

AI	artificial intelligence
CE	common era
EFL	English as a foreign language
EIC	East India Company
FICCI	Federation of Indian Chambers of Commerce and Industry
GCI	global competitiveness index
GDP	gross domestic product
HDI	human development index
IHCL	Indian Hotels Company Limited
IMF	International Monetary Fund
SFL	systemic functional linguistics
TTCI	travel and tourism competitiveness index
UNESCO	United Nations Educational, Scientific, and Cultural Organization
VR	virtual reality

Acknowledgments

This book is a product of the vision and cumulative efforts of the editorial board and technical reviewers. This interdisciplinary initiative attempts to offer a space for intellectual deliberations on language in cross-cultural communication in travel and tourism.

We extend our sincere gratitude to all the authors who have contributed to this volume with their scholarly and intellectually stimulating discussions. We also thank the staff at Apple Academic Press for their unflinching support and consistent cooperation.

Foreword

In the bustling realm of travel and tourism, where every journey is a story waiting to be told, the book you hold in your hands stands as a beacon of insight and guidance amidst the tumultuous seas of our global economy. *Language and Cross-Cultural Communication in Travel and Tourism: Strategic Adaptations* emerges not just as a compendium of scholarly research, but as a roadmap for stakeholders navigating the intricate pathways of the tourism industry, seeking to steer toward sustainability and success. Contained within these pages is a tapestry woven from the collective wisdom of scholars spanning the globe, each offering their unique insights to untangle the intricate web of challenges and opportunities inherent in this multifaceted field. Through the lens of language and cross-cultural communication, this book illuminates the diverse landscapes and cultures that shape the tourism experience. As we embark on this intellectual journey, it becomes evident that the obstacles we face are as varied as the destinations we explore. Yet, armed with the knowledge and guidance provided within these chapters, we are empowered to navigate these complexities with clarity and purpose, charting a course toward a future where travel and tourism serve as catalysts for understanding and connection across borders.

Chapter 1 invites us to contemplate the essence of passenger identity through the lens of public transport awareness campaigns. From the vibrant posters adorning railway stations in Japan to the instructive illustrations gracing those in France, we witness the power of visual communication to shape behavioral norms and societal expectations, revealing the intricate tapestry of cultural nuances that define our interactions. In Chapter 2, the philosophical underpinnings of tourism intersect with the mosaic of cultural diversity, prompting reflection on the transformative potential of multicultural encounters. Here, amidst the allure of exotic destinations, lies the crucible of economic growth and social cohesion, where the richness of human experience converges with the imperatives of sustainable development. The COVID-19 pandemic takes center stage in Chapter 3, illuminating the evolving landscape of intercultural communication in tourism. Through the virtual realms of social media, we embark on a journey of exploration and connection, discovering how digital influencers shape perceptions and beckon travelers to distant shores, even amidst times of uncertainty.

As we navigate the turbulent waters of contemporary challenges, Chapter 4 beckons us to confront the intersection of tradition and modernity, as exemplified by the Durga Puja festivities in Kolkata. Here, amidst the vibrant tapestry of cultural expression, we witness the resilience of the human spirit, as communities adapt and evolve in response to the exigencies of our time. Chapter 5 offers a poignant reflection on the pandemic's impact on over-tourism, heralding a paradigm shift towards eco-friendly and responsible travel practices. Amidst the upheaval, we discern the seeds of opportunity, as policymakers and academics unite in reimagining a future, where sustainability reigns supreme. In Chapter 6, the echoes of history resonate in the hallowed grounds of Gallipoli, inviting us to traverse the hermeneutical bridge of empathy and understanding. Here, amidst the scars of conflict, lies the potential for reconciliation and healing, as we embrace the stories of others as our own.

Literary ethnography takes center stage in Chapter 7, as we embark on a journey into the heart of the Sundarbans, where the rhythms of life intersect with the imperatives of ecological stewardship. Through the lens of fieldwork and lived experience, we confront the complexities of cultural negotiation and adaptation in the face of modernity. Chapter 8 beckons us to explore the hidden gems of India's rural landscapes, where the warmth of hospitality meets the allure of uncharted territories. Here, amidst the chai stalls and winding lanes, lies the promise of discovery and connection, beckoning travelers to venture beyond the beaten path. The early tourist industry in Alaska takes center stage in Chapter 9, as we unravel the complexities of indigenous representation and commodification. Amidst the rugged beauty of the Last Frontier, we confront the tensions between economic growth and cultural preservation, as we strive to forge a path towards sustainable tourism practices.

Chapter 10 offers a linguistic and visual journey into tourism advertising, where words and images converge to weave tales of allure and enchantment. Through the prism of Vietnam's cultural landscape, we uncover the power of storytelling to captivate the imagination and beckon travelers to embark on their odyssey. In Chapter 11, we trace the evolution of cultural and heritage tourism in Kochi, where the echoes of the past resonate amidst the clamor of urbanization. Here, amidst the labyrinthine

Foreword   xix

streets and ancient landmarks, lies the challenge of balancing economic development with the preservation of cultural heritage. Chapter 12 invites us to contemplate the promise and potential of rural tourism in India, where the verdant landscapes and vibrant communities beckon travelers to embark on a journey of discovery and connection. Here, amidst the rustic charm of village life, lies the opportunity to forge meaningful connections and foster sustainable development. Finally, in Chapter 13, we confront the transformative power of social media in shaping the tourism landscape, as digital platforms emerge as the vanguard of promotion and engagement. Through the lens of modern electronic techniques, we explore the myriad ways in which social networking sites fuel the imagination and inspire wanderlust.

I firmly hold the conviction that the themes explored within these chapters are of paramount significance. They have been meticulously researched and discussed, offering valuable insights into various facets of the travel and tourism industry. I extend my heartfelt commendation to the dedicated editors and authors whose tireless efforts have culminated in this insightful publication. Their commitment to addressing crucial issues impacting our economy is commendable, and this book serves as a vital resource for enhancing our understanding of these complex challenges.

—Dr. Dhanonjoy Kumar
Professor
Department of Management
Islamic University, Kushtia, Bangladesh

Preface

This volume seeks to provide myriad scholarly insights into the linguistic, social, and cultural phenomena that are embedded and engendered in the context of tourism. The chapters in this book incorporate global case studies and demonstrate the application of theoretical discourses rooted in various disciplines such as marketing, management processes, literature, media studies, and linguistics. They also address the repercussions of the COVID-19 pandemic on the tourism industry and the simultaneous rise in the influence of social media in the cultural semioscape of tourism.

With its tripartite focus on language, tourism, and cross-cultural aspects, this book strives to make fresh scholarly interventions in the field and will assist students, scholars, and educators in cutting across diverse knowledge domains.

Introduction

As one of the rapidly evolving industries in the world, tourism has transcended its immediate economic concerns and has become a major signifier of cultural patterns and cross-cultural communications. Theoretical and scholarly undertakings in tourism have steadily grown, bringing a wide range of discourses and ideologies within its ambit. With this, the function of language has become the subject of scrutiny in the context of intellectual deliberation vis-à-vis travel and tourism.

One of the major concerns in examining travel and tourism is understanding the communication patterns that are engendered in the process. Such patterns, forged by diverse communication practices, call for an examination of the sociocultural and ethnographic context of the participants in such communicative events. While a person's long-term residence and/or place of birth determine whether they belong to a particular speech community, a discourse community is created if it satisfies a number of criteria, such as:

- It adheres to the goals that have been established by the public;
- It has a system in place for communication among its members that includes both the sharing of information and the solicitation of feedback
- This structure is secured by one or more genres
- This framework has a developed vocabulary
- It has the intention of expanding its membership (Swales, 1990: 24–27).[1]

Similarly, a tourism discourse community may be identified as a large community that communicates shared perceptions and beliefs through the aforementioned communicative loop.

The intricate connections between cultural variables, environmental circumstances, knowledge, and emotion lead to the communicative

---

[1] Swales, J. (1990). *Genre Analysis: English in Academic and Research Settings.* Cambridge: Cambridge Applied Linguistics.

practices of people. Contextual factors, such as attitudes, values, and customs of an ethnic group, are developed and disseminated through communication, rendering the process of communication its significance. Cross-cultural interactions are referred to as communicative settings that involve people from divergent cultures; such interactions often result in miscommunications owing to the heterogeneity of linguistic content and markers of respective cultures.

Culture is defined by anthropologists as "the collective programming of the mind by which members of a group or social category are recognized from others" (Hofstede 2012).[2] Hofstede, in his anthropological theory of culture, refers to it as the "software of the mind." Cross-cultural communication studies recognize how culture shapes our behavior, thought process, communication, and linguistic practices (Dodd 1995).[3] Cultural variations contribute to everyone's unique communication preferences, worldviews, and personalities, which in turn produce divergent consequences. The following four types adequately capture the entire notion of symbols from the wide variety of names used to define them as manifestations of culture (Albu, 2015).

Symbols: Words, gestures, images, and other things that have a particular meaning and can be understood by others outside of the society that created them are known as symbols. This category also includes slang words and words from a language. The symbols were positioned outside in the direction of the surface layer for this reason.

Heroes: They are individuals, alive or dead, real or imagined, who possess the attributes valued in a culture and act as models for behavior.

Rituals: Even though they do not directly contribute to the practical achievement of desired goals, they are collective activities that, within a culture, are regarded as crucial to the social plan. Consequently, their completion is valuable in and of itself.

Values: Values are the fundamental components of culture; they are broad propensities to favor some conditions over others. These emotions are bipolar, having both a positive and a negative side.

Practices: These consist of ceremonies, heroes, and symbols. These are obvious to an outsider, yet their cultural value is hidden and solely depends on how the group members themselves understand them.

---

[2] Hofstede, G. (2012). *Cultures and Organizations. Mental Software.* Bucharest: Humanitas.

[3] Dodd, C. (1995). *Dynamics of Intercultural Communication.* Dubuque, Iowa: William C. Brown Company Publishers.

Introduction   xxv

In his book *Elements of Semiology*, Roland Barthes contends, "dès qu'il y a société, tout usage est converti en signe de cet usage" [once society exists, every usage is converted into a sign of this usage]. If it were not for the tourism industry, which serves as an excellent example and can offer significant guidance and illumination, the idea that usage can become a sign in and of itself might remain somewhat ambiguous and provide the analyst with little methodological instruction in how to pierce alibis and what to look for. The justifications a culture makes to repurpose its customs are of no relevance to the tourist. The traveler is interested in everything as a reflection of themselves and as an illustration of a common cultural custom. In this point, one can consider the Anglo-Indian culture of Kolkata's *Bow Barracks*. Along with Chinese, Bengali, and Gujarati residents, Anglo-Indian families also reside there. It is a cosmopolitan area that serves as a melting pot for various cultures. They demonstrate Jean Baudrillard's contention that signifier, not wants or use-value, should serve as the foundation of any valid theory of social objects.

A potent semiotic operator in the tourism industry is the connection between language and cross-cultural communication. Both the language itself and the discourse on tourism depend on the notion of understanding the real tale and how the structure works. In this sense, language and communication in travel are seen as indicators of language usage, and a big part of tourism is the search for these indicators. The 13 chapters on this issue clearly explain the potential contributions that scholars with interests in cross-cultural communication, linguistic narration, and linguistic tourism can make to this field of study.

The chapters in this volume offer a diverse range of perspectives on the notions of culture and communication, ranging from intersections in travel and tourism to intersections in travel literature to intersections in the "semioscape" of tourism. The chapters have been grouped into three sections, that is, (i) *Cross-Cultural Communication in Travel and Tourism*, (ii) *Narrative of Place and Space*, and (iii) *Travel and Tourism Industry*, so that they roughly fit into one of these dimensions.

In this context, the first chapter of Kurihara et al. discusses how passengers are portrayed in public transportation awareness program posters. The writers' choice of the cross-cultural area between Japan and France serves this objective. This gives them the chance to compare the awareness campaigns in Japan and France while also learning more about how both nations' public transportation users interpret and portray the reality of these campaigns at the intra- and intercultural levels of communication. The differences in communication styles often create an identity. Culture, in this context, is often identified and associated with the creation of a nationwide group that has emerged to retain the customs of this newly emerged nation-state.

The second chapter of this book follows the same exact course as the first, as S. K. Maurya seeks to understand how multicultural and multireligious countries develop their distinctive features for their own tourists. There is a widely held belief, according to the author, that cultural or ethnic fractionalization will inevitably lead to difficulties in communication and cooperation, which will lead to less favorable economic growth, less resilient social and economic systems, and ultimately, a reduction in productive capacity. The ensuing conflicts and difficulties may deter travelers, endangering the hospitality sector across cultures. On either side, cultural fractionalization and the resulting cultural diversity might be viewed as valuable advantages since they provide a varied range of knowledge, practices, skills, and traditions that can foster creativity and innovation.

Saha and Raghuvanshi in Chapter 3, Datta in Chapter 4, and Deb, and Kumar in Chapter 5 discuss the effects of COVID-19 on the overall tourism sector. Saha and Raghuvanshi point out that during the pandemic, social media and social networking sites played a significant role in pushing advertising campaigns in every industry, including tourism. They are discussing intercultural communication in tourism and its connection to social media. In addition to Incredible India's conventional tourist promotion strategy, numerous travel blogger users also use trip films to advertise the destination. In this regard, their study investigates this topic by looking at various travel bloggers on Facebook and how they utilize their posts, photographs, and videos to persuade people to visit the destination. On the other side, Deb and Kumar discuss how this pandemic has negatively impacted the travel and tourism sector. The authors' opinions on the contribution of the coronavirus to the cessation of mainstream tourism operations are shown in Chapter 5. It also tries to explain how academics and policymakers may play a crucial part in the current issue by rethinking and redesigning the current curriculum and regulations to prepare future generations to adopt more environmental friendly and responsible travel and tourist practices. The fourth chapter is a little bit different than the first two. Here, the author examines the concept of sacred and profane

Introduction   xxvii

from the perspective of COVID-19 and takes into account the Durga puja in Kolkata, one of UNESCO's intangible cultural heritages of humanity. By putting the theme of puja's inertia in its proper context, the author dissected the aforementioned ideas in the manner of Emile Durkheim.

An important recurring subject in travel narratives is the observation and interpretation of spatial relationships and geographical features. The assumption made in Part 2 of the book, *Narrative of Place and Space*, is that the said two can be viewed from a wide range of perspectives and levels, revealing meaningful and semiotically specified sets of linkages. The section contains three chapters with this idea in mind, and it starts with Çöker's *Transfiguring the Troubled Past in Gallipoli: A Woman's Perspective* in Chapter 6. Through utilizing the viewpoint of French philosopher Paul Ricoeur, this chapter makes the case that ethical engagement is achievable by creating a hermeneutical bridge. It further employs the technique of retelling historical narratives by exchanging stories in order to interpret otherness by offering the opportunity to view one another from the perspective of the other.

Chapter 7 takes a step further toward the discussion of narrative space by following Rakshit's *Literary Ethnography and Travel Aesthetics*, Amitav Ghosh's *The Hungry Tide*, and *Jungle Nama: A Story of the Sundarban*. The chapter therefore begins by locating Puri and Castillo's call for "theorising fieldwork in the humanities" as a key factor in why travel tales in postcolonial countries include the lived experiences of local people as residents and the writer as a literary ethnographer. The chapter will then demonstrate how renowned Indian novelist Amitav Ghosh, working across a range of geographic contexts, historicizes the tragic fate of the people of the Sundarbans and their embodied experiences in his two works, *The Hungry Tide* (2004) and *Jungle Nama: A Story of the Sundarban* (2021), drawing on recent scholarship on "literary travel" that strategically adapts the fieldwork narrative and literary narrative through the regional travel aesthetics in the postcolonial studies.

Chapter 8 illustrates travel accounts using a geolocative premise. In their work, *Journeying Through the Lesser-Known Indian Spaces: A Reading of Bishwanath Ghosh's Chai Chai*, Pati and Longkumer make the argument that a traveler will learn more about a certain country or location (in their case, India) by visiting little pockets rather than well-known locations. They claim that there is a recent spike of tourists who want to see these lesser-known regions of a country that are frequently overshadowed by well-known tourist attractions. This chapter, which follows the plot of Biswanath Ghosh's book *Chai, Chai: Travels in Places Where You Stop But Never Get Off*, takes us through the lanes and alleys of India in this regard.

We are all aware that the tourism sector refers to a group of businesses that enable travel for a variety of objectives, including pleasure and business, by providing the necessary infrastructure, goods, and services. The final half of the book will revisit this idea of the travel industry while focusing on places like Alaska, Kochi, and most importantly on rural tourism. This section starts with Parham and Williams's *Imagined Communities: The Development of the Early Tourist Industry in Alaska and the Marketing of the Indigenous Experience*, in Chapter 9. Through the analysis of curio stores, cruise line promotions, and published diaries of Alaska visitors, this work focuses on the development of the cruise business in Southeast Alaska from 1870 to 1940 and the emergence of the curio trade to enhance tourist traffic to the area.

In Chapter 10, Loan, in his work, *Multimodal Approach to Tourism Advertising Discourses*, focuses on how Vietnamese copywriters integrate language and picture to generate meaning for tourism advertising discourses using theoretical frameworks such as Martin and White's appraisal framework, Kress and van Leeuwen's visual grammar, and Bhatia's move structure. The work has also shown that Vietnamese copywriters employed grammar and lexis to reflect how they felt about the discourse used in tourism promotion. According to the findings, copywriters arrange travel commercials to communicate three different types of meaning based on visual grammar analysis, including representation, interaction, and composition.

The heritage tourism sector is the main topic of Chapter 11 by Munoth et al., with a concentration on Kochi. In their chapter, *The Saga of Kochi: Cultural and Heritage Tourism Overview*, Munoth et al. make an effort to chart the evolution of Kochi's unique culture and heritage as well as changes to the city's urban landscape over time. The global tourism business includes rural tourism as a significant component. Many locations rely on their rural tourism offerings to bring in much-needed funds for the local economy.

Given that it is inextricably related to green environments and frequently eco-friendly kinds of travel like hiking and camping, it is intimately aligned with the idea of sustainable travel. Following this tourist

Introduction

xxix

route, Chapter 12 of the book, through the work of Kumar et al., *Rural Tourism in India: Constraints and Opportunities*, tries to review several studies and will offer significant views on the challenges and opportunities of rural tourism in India, along with some recommendations for policy-makers on how to enhance their efforts to promote rural tourism.

In particular, information search and decision-making behaviors, tourism promotion, and the focus on best practices for engaging with customers are all significantly influenced by social media. Utilizing social media to advertise travel-related products has shown to be a successful tactic. Bhatt and Verma's work in Chapter 13 has translated this concept.

CHAPTER 1

Passenger Identity in Public Transport Awareness Campaign Posters: Contrastive Study of Communication Styles in Japan and France

YUI KURIHARA[1], JUNGAH CHOI[2], and YOSHINORI NISHIJIMA[3]

[1]*Graduate School of Humanities, Osaka University, Minoh, Japan*

[2]*Department of International Communication, Nagasaki University of Foreign Studies, Nagasaki, Japan*

[3]*Institute of Human and Social Sciences, Kanazawa University, Kanazawa, Japan*

ABSTRACT

We often see posters on trains and at railway stations. Certain posters provide passengers with travel-related information, whereas others instruct them on appropriate behavior on trains. Such awareness campaign posters vary across countries in terms of design and expression. In Japan, posters are often accompanied by illustrations of mascots; in France, few posters display mascots. The purpose of this study is threefold: (1) to classify awareness campaign posters into four types of communication styles based on mono- and dual-communication styles according to Yamaoka's (2005) literary communication model; (2) to compare the communication styles of the Japanese and French awareness campaigns; and (3) to identify how passengers are perceived and depicted in posters in Japan and France. The differences in communication styles reveal how each country defines the identity of its passengers.

Language and Cross-Cultural Communication in Travel and Tourism: Strategic Adaptations.
Soumya Sankar Ghosh, Debanjali Roy, Tanmoy Putatunda, & Nilanjan Ray (Eds.)
© 2025 Apple Academic Press, Inc. Co-published with CRC Press (Taylor & Francis)

1.1 INTRODUCTION

In the past, various signs posted in public spaces, such as train stations, have received attention from the perspective of linguistic landscapes, and have been analyzed mainly in terms of the character types and multilingual notation of the linguistic expressions displayed. Among such public signs, this study focuses on awareness campaign posters displayed at railway stations and on trains. These posters are intended to effectively convey diverse information: from the prohibition of dangerous behavior to instructions regarding manners. In other words, such posters can be analyzed from the viewpoint of communication, in which the sender effectively conveys a message to the receiver.

However, the method of communication may differ across cultures. The purpose of this study is to analyze awareness campaign posters in Japan and France using a communication model, and to clarify the differences in message transmission strategies between the two countries. First, in Sections 1.2 and 1.3, a communication model is proposed such that posters between different languages can be appropriately compared. Next, in Sections 1.4 and 1.5, we analyze Japanese and French manner posters, respectively, using the common framework, and compare the results in Section 1.6.

1.2 DISCUSSIONS OF FRAMEWORKS FOR ANALYSIS

1.2.1 COMMUNICATION MODEL

Several studies have been conducted on awareness campaign posters, such as Yamauchi (2004), Mizuta (2013), Fujimura and Taniguchi (2017), Bayne (2018), and Yamamoto (2020); however, few studies have been performed based on the communication framework of sender and receiver. Manners awareness posters are public advertisements, which are argumentative discourses that encourage or forbid certain behaviors toward their target audience. However, such advertisements may strategically create an imaginary world that is different from the argumentative discourse field and send messages from within this space. Such a scheme in which another imaginary discourse exists within one discourse is similar to the scheme in which the discourse is performed by the characters in the story related by a narrator. In fact, Wang (2020) has already analyzed movie posters by applying Yamaoka's (2005) model of communicating subjects in narratives, revealing that in multilayered film advertising posters, the relationship between the sender and receiver of a message is not as simple as that between the "sender and receiver of a message" within the story. Choi (2021) compared the results of the analysis of movie posters by Wang (2020) with those of her analysis of the Japanese awareness campaign posters and found that the awareness posters that warn against annoying behavior often show a more complex style involving characters. Based on the results, Choi (2021) indicated that this complex style is double-structured to indirectly convey a message to the reader, which is characteristic of the communication style in Japanese society.

The purpose of this study is to analyze awareness campaign posters in Japan and France, which include diverse content: from the prohibition of dangerous behaviors to instructions regarding manners, using a common communication model, and to clarify the differences in message transmission strategies between the two countries. To propose a communication model that enables this analysis, Section 1.2.2 critically examines Yamaoka's (2005) model for analyzing communicating subjects in narratives. Then, in Section 1.2.3, we modify the model and propose a communication model in which Japanese and French awareness campaign posters can be appropriately compared.

1.2.2 YAMAOKA'S (2005) COMMUNICATION MODEL

Based on Adams' (1985, p. 12) communication model shown in (1) below, Yamaoka (2005, pp. 43–44) proposed an analytical model, as shown in (2), for the analysis of the complex relationship between subjects involving a narrator, a hearer, and characters in a story.

(1) W→(S→text→H)→R (W: writer, S: narrator, text: content, H: hearer, R: reader)

(2) W→(S1→(S2→text→H2)→H1)→R
(W: writer; S: narrator; text: content; H1: hearer; R: reader; S2: character1; H2: character2 or character1)

The text was sent from the left and transmitted to the right. The parentheses represent the worlds in which the discourse was constructed. The outer parenthesis is the world in which the narrative discourse is constructed, and the inner parenthesis is the narrated world, that is, the world in which the characters construct the discourse. The writer, W, and the actual reader, R, are defined as subjects located outside the parentheses and not involved in the construction of any type of discourse. The fictional narrator, S, who is responsible for the narrative discourse as a device or function, conveys messages by himself/herself and/or through the discourse of the characters.

Yamaoka's model, in which the omnipresent existence of the narrator is acknowledged, cannot fully explain, for example, the problem of the existence or absence of a narrator in the story written in the narrative past ("*passé simple*") and in the third person in French, or the differences in various styles of reported speech. However, it is an appropriate model for describing the complex relationships between senders and receivers in cases where the discourse is constructed on multiple levels, including a fictional discourse. Furthermore, in the study of awareness campaign posters, it is important as a communication strategy for how the recipients are designed as well as who is sending the message. Therefore, Yamaoka's model was adopted for our analysis.

1.2.3 MODIFICATION OF YAMAOKA'S (2005) COMMUNICATION MODEL

Moving the topic from narratives to posters, the levels of discourse can be redefined as follows:

(3) Outer parenthesis: level of narrative discourse → level in poster discourse.
Inner parenthesis: level of the narrated world (the world in the story) → level of the fictional world (discourse created in poster discourse).

The poster discourse level is that where the awareness campaign message is conveyed directly to the targets of the poster, so deictic expressions such as "here" or "now" are understood at this level, that is, concerning the "here and now" of the targets. The imaginary world level is not directly related to the targets of the poster, and the deictic terms are interpreted in relation to the "here" or "now" of the fictional world—without any reference to the situation of the targets. Therefore, Yamaoka's model for narrative analysis can be revised to Model (4) when transposed to the context of awareness campaign posters.

Passenger Identity in Public Transport Awareness Campaign Posters 5

(4) W→(S1→(S2→text→H2)→H1)→R
(W: advertiser; S1: narrating function of poster discourse; text: poster content; H1: target of poster; R: reader; S2: character in an imaginary world within the poster discourse; H2: character in an imaginary world within the poster discourse).

Certainly, there is also an awareness campaign poster (e.g., Figure 1.1) with a structure without an imaginary world level, as in (1) above. When communication model (1) is transposed into the context of the awareness campaign poster, as shown in Figure 1.1, it is described as the model in (5).

FIGURE 1.1[1]
(*Source*: Reprinted with permission from SEMITAG, 2019 © 2019 SEMITAG -Realization Hula Hoop communication agency. Grenoble, France)
S1: Les journées du bon sens :)
"Common sense days"

(5) W→(S1→text→H1)→R
(W: advertiser; S1: narrating function of poster discourse; text: content; H1: target of poster; R: reader).

There are explicit and implicit criteria for determining the existence of a fictional world level within a poster.

A) Explicit criterion: Presentation of concrete scenes using images (Figure 1.2).

B) Implicit criterion: Existence of utterances that do not belong to the "now and here" of the poster discourse level (Figure 1.3).

---

[1]Hereinafter, the poster discourse level is depicted by a black line. It matches the outer circumference of the poster.

6 Language and Cross-Cultural Communication in Travel and Tourism

FIGURE 1.2[2]

(*Source*: Reprinted with permission from Metro Cultural Foundation, 2019 © 2019 Metro Cultural Foundation, Tokyo, Japan)
S1: Block ?
Densha-no noriori, mawari-no hito-o block-shitenai?
"Block? Aren't you blocking the passengers around you when they get on and off the train?"

FIGURE 1.3

(*Source*: Reprinted with permission from Tisséo, 2020 © 2020 Tisséo – Realization VERYWELL agency. Toulouse, France)
S2: Et oui, même pour 3 min de métro il faut un ticket.
"Yes, even for a 3-minute ride, you need a ticket."

In Figure 1.2, a scene in a fictional train is depicted in the white frame, indicating a different world from that of the poster discourse. In Figure 1.3, *Et oui, même pour 3 min de Métro il faut un ticket* ("Yes, even for a 3-minute ride, you need a ticket") is a message uttered by a controller responding to a fraudster in a concrete scene. It does not refer to the "here and now" of the target of the poster discourse level, where the poster discourse is performed, but introduces a world different from that of the poster discourse.

Based on the above criteria, this chapter focuses on whether the utterance in a poster is attributed to S1, which is located at the poster discourse level, or to S2, which is at the imaginary world level and analyzes posters. Defining the origin of the utterance enables us to analyze the receiver's side as well, but because of the complexity of the analysis and space limitations, we will not delve into the receiver side in this analysis.

1.3 METHODS FOR ANALYSIS

1.3.1 CRITERIA FOR DETERMINING THE ATTRIBUTION/ORIGIN OF UTTERANCES

In the same way that utterances in a narrative can be attributed either to the narrator or to the characters, utterances in a poster can be attributed to subject

---

[2] An area surrounded by a white line belongs to the imaginary world level.

S1 in the poster discourse level and/or to subject S2 in the imaginary world level. The criteria for determining the attribution of utterances were as follows.

1.3.1.1 CRITERIA FOR IDENTIFYING S1'S UTTERANCES AT THE POSTER DISCOURSE LEVEL

There are explicit and implicit criteria determining S1's utterances at the poster discourse level.

A) Explicit criterion: The use of expressions addressed directly to the interlocutor, that is, the target H of the poster discourse level, such as the pronoun "you" and the use of imperative and invitation forms (Figure 1.4).

B) Implicit criterion: The use of meta-expressions that indicate the entities of the imaginary world level with nouns or third-person pronouns (Figure 1.5).

FIGURE 1.4

(*Source*: Reprinted with permission from Bureau of Transportation, 2015 © 2015 Bureau of Trans-porta-tion. Tokyo Metropolitan Govern-ment, Japan
S1: Tenimotsu-wa mawari-eno hairyo-o wasurezu-ni.
"Please be considerate of others when handling baggage." (upper side)
S2: Komarimasu! Anata-no nimotsu, shirampuri!!
"How annoying! When you pretend ignorance your bag!!" (bottom right)

FIGURE 1.5

(*Source*: Reprinted with permission from CITURA, 2018 © 2018 CITURA (TRANSDEV). Reims, France)
S2: Dans le bus & le tram dès le départ je me tiens bien.
"In the bus & the streetcar from the start I hold well on the pole"
S1: Un cornichon ça fait pourtant toujours attention!
"Even a pickle (=idiot) is always careful!"

In Figure 1.4, after giving situational descriptions from *Komarimasu! Anata-no nimotsu, shirampuri!!* ("How annoying! When you pretend ignorance your bag!!"), the imperative form *hairyo-o wasurezu-ni!* ("Please, don't forget to be considerate of", i.e., "Please be considerate of") is used and requests the target H of the poster discourse level to confirm and cooperate. In Figure 1.5, the subject noun phrase *Un cornichon* ("a pickle") in the 2nd utterance (which is in the square below) metalinguistically designates a passenger (Mr. Idiot) in a fictional streetcar, and it can be said that the utterance is made by the subject not located in the imaginary world, that is, subject S1 of the poster discourse level.

1.3.1.2 CRITERIA FOR IDENTIFYING S2'S UTTERANCES AT THE IMAGINARY WORLD LEVEL

There are also explicit and implicit criteria for S2's utterances at the imaginary world level. Both the explicit and implicit criteria were categorized in more detail.

A) Explicit criterion 1: The text is presented with an image of the subject speaking at the imaginary world level (Figure 1.6).

B) Explicit criterion 2: Although there is no image of the speaking subject, the text is presented as an utterance of the subject of the imaginary world level with speech bubbles, quotation marks, etc. (Figure 1.7).

FIGURE 1.6

(*Source*: Reprinted with permission from Metro Cultural Foundation, 2012 © 2012 Metro Cultural Foundation, Tokyo, Japan
S2: Naze iyahon-no otomore-ni kizukanai-no?
"How come you don't notice your earphones leaking?"

FIGURE 1.7

(*Source*: Reprinted with permission from TBM, 2018 © 2018 TBM (KEOLIS BORDEAUX METROPOLE), Bordeaux, France)
S2: "J'ai déjà dit non!"
"I already said no!"
S2': "Allez! donne ton 06..."
"Come on! give your number 06..."

In Figure 1.6, a character (mascot) on the bothered person's shoulder is mumbling on behalf of him/her, looking at the bothering man as well as the bothered person. Figure 1.7 provides a dialogue between a stalker and his victim. The speech bubbles clearly show that each sentence is spoken by the man and the woman in the image, respectively.

A) Implicit criterion 1: The content contains modalities such as subjective judgments, questions, and exclamations that can only be attributed to the subject (person or mascot) of the imaginary world level (Figure 1.8).

B) Implicit criterion 2: Containing direct expressions (deixis) that originate from the imaginary world level (Figure 1.9).

C) Implicit criterion 3: Use of anacoluthon, broken colloquial expressions, interjections, "yakuwari-go" ("stereotypical utterances of characters"), handwritten characters, etc., and also from the content, "a specific speaker who does not fit in the poster discourse level" can be identified (Figures 1.10 and 1.11).

FIGURE 1.8

(*Source*: Reprinted with permission from Metro Cultural Foundation, 2014 © 2014 Metro Cultural Foundation. Tokyo, Japan)
S2: Pipip! Mawari-ni oto-ga morete-iru-yo
"Beep Beep! Sound is leaking around."

FIGURE 1.9

(*Source*: Reprinted with permission from TBM, 2013 © 2013 TBM (KEOLIS BORDEAUX METROPOLE). Bordeaux, France)
S2: Je monte Je valide.
"I board I validate my ticket."

FIGURE 1.10
(=Figure 1.3)
S2: Et oui, même pour 3 min de métro il faut un ticket.
"Yes, even for a 3-minute ride, you need a ticket."

FIGURE 1.11
(*Source*: Reprinted with permission from Bureau of Trans-portation, 2017 © 2017 Bureau of Trans-portation. Tokyo Metropolitan Government, Japan)
S1: Kakekomi taigā.
"A last-minute rush tiger"

These implicit criteria are often consistent with the criteria for identifying free indirect speech (Banfield, 1995, pp. 71–75; Mikame, 2017). In Figure 1.8, subjective judgment is expressed in the form of warning passengers who are listening to music with loud sounds without regard for their surroundings. In Figure 1.9, next to the image of the llama with a boarding card in its mouth, it is written: *Je monte Je valide* ("I board I validate"). The subject (*Je*) boards in an imaginary world, and this utterance does not belong to S1 of the poster discourse level. Figure 1.10, which begins with *Et oui* ("Yes, that's right"), can be considered an imaginary conversation between a fraudster and a conductor and cannot be attributed to S1 at the poster discourse level. In Figure 1.11, S1 uses a *dajare* ("a homophonic pun" i.e., *Kakekomi taigā* is almost the same sound as want to rush into because *tai* of the word *taigā* means "wish/want") to explain that the passenger in an imaginary world, who wears a tiger's headgear.

1.3.2 BASIC PATTERNS OF COMBINATIONS OF LEVELS AND SUBJECTS

When we analyze the combinations of the discourse level and the subject attributed to the utterance based on the above criteria, the basic patterns are as follows.

1.3.2.1 PATTERN 1: W→ (S1→TEXT→H1)R

The poster is constructed only at the poster discourse level. Naturally, the utterance on the poster is attributed to S1.

Passenger Identity in Public Transport Awareness Campaign Posters 11

FIGURE 1.12
(=Figure 1.1)
S1: Les journées du bon sens :)
"Common sense days"

FIGURE 1.13
(A collaboration by railway operators, 2017, Japan)
S1: Yamemashō, aruki sumaho.
"Stop using your phone while walking."

In Figure 1.12, only the message is displayed. The footprints in Figure 1.13 are the images of the message *aruki sumaho* (commonly referred to *as texting while walking*), and no imaginary world has been created. Additionally, the utterance is attributed to S1 at the poster discourse level.

1.3.2.2 PATTERN 2: W→ (S1→ TEXT→ H1) R

The imaginary world is created in the poster by the image, but the utterance in the poster belongs to S1. The communication model is consequently the same as in Pattern 1, but the text often refers to the imaginary world level.

FIGURE 1.14
(*Source*: Reprinted with permission from Tub, 2021 © 2021 Tub – Realization Studio Lannion, Saint-Brieuc, France)
S1: Mettez vos déchets au panier pour davantage de propreté.
"Put your garbage in the basket for more cleanliness."

FIGURE 1.15
(*Source*: Reprinted with permission from Metro Cultural Foundation, 2019 © 2019 Metro Cultural Foundation. Tokyo, Japan)
S1: Space?
Anata hitori-de, supēsu-o torisugite-imasenka?
"Space?
Aren't you taking up too much space for yourself?"

Figures 1.14 and 1.15 depict a basketball player scoring a goal and a man spreading his legs inside a car in an imaginary world, respectively. The utterance in Figure 1.14 is formulated in the imperative form and that in Figure 1.15 in the interrogative form with the second-person pronoun *Anata* ("you"). This implies that in both cases, each utterance is addressed directly to the target, located at the poster discourse level. These two messages are thus utterances by S1 at the poster discourse level.

1.3.2.3   PATTERN 3: W→ (S1→ (S2→ TEXT→ H2) H1)→ R

The utterance belongs to the subject S2 located at the imaginary world level.

FIGURE 1.16

(*Source*: Reprinted with permission from Bibus Brest, 2020 © 2020 Bibus Brest. Brest, France)
S2: Vendredi 19h30, je vais à un spectacle au Quartz.
Je fais attention au tram.
J'arrive sans accroc!
"Friday 7:30 pm, I go to a show at the Quartz. I pay attention to the streetcar. I arrive without a hitch!"
No corresponding example in Japan.

Figure 1.16 shows a woman wearing sunglasses and earphones. The sunglasses show the streetcar and theater, and the utterance is constructed in the first-person singular *je*. From these points, it can be said that the subject *je* ("I") who goes out to the Quartz (theater in Brest) while being careful of the streetcar is the woman represented in this image, that is, S2 positioned in the imaginary world of "Vendredi 19h30." There is no corresponding example in Japan.

1.3.2.4   PATTERN 4: W→ (S1→ TEXT→ H1)→ R AND W→ (S1→ (S2→ TEXT→ H2) H1)→ R

In combination with Patterns 2 and 3 above, multiple utterances are present, which belong to either S1 or S2.

Passenger Identity in Public Transport Awareness Campaign Posters 13

FIGURE 1.17

(*Source*: Reprinted with permission from TaM, 2017 © 2017 TaM – Realization KFH. Montpellier, France)
S2: SVP (=S'il vous plait)
"Please"
S1: Confortable pour les uns, indispensable pour les autres.
"Comfortable for some, indispensable for others."

FIGURE 1.18

(*Source*: Reprinted with permission from Bureau of Transportation 2020 © 2020 Bureau of Transpo-rtation. Tokyo Metro-politan Government, Japan)
S2: Hashiranai-de!
"Don't run!"
S1: Kakekomi jōsha-wa yamemashō.
"Please do not rush to board the train."

Figure 1.17 describes a scene in the train car where a pregnant woman says "please" to another woman sitting on the bench. The scene takes place within a white circle, and outside this circle is a message that adds an objective explanation to this scene. The latter message is attributed to S1 at the poster discourse level that does not belong to the fictional world, and the message is transmitted from the two levels. In Figure 1.18, a Japanese folk tale character named Momotarō tries to stop the animals that also appear in the story from rushing into the car. Below the same, a polite message to not rush into the car is conveyed—an instruction thought to belong to S1 of the poster discourse level. At this discourse level, explanations in foreign languages such as English, Chinese, and Korean are often added in the case of Japan. Figure 1.19 shows an image of the structure of Pattern 4, composed of these two levels.

```
Poster discourse level
        W → (S1 → text → H1) → R
    Imaginary world level
        W → (S1 → (S2 → text → H2) → H1) → R
```

FIGURE 1.19 An image of dual structure

Pattern 4 shows that the utterance by S1 of the outer space of the poster discourse level and the utterance by S2 of the inner space of the imaginary world coexist.

In this chapter, we call Pattern 4, as shown in Figure 1.19, a dual communication structure and Patterns 1 to 3, in which only one of the two levels produces an utterance, a mono communication structure.

1.4 ANALYSIS OF AWARENESS CAMPAIGN POSTERS IN JAPAN

1.4.1 MATERIALS COLLECTED

We collected two types of awareness campaign posters from 2011 to 2020: (1) those produced by eight individual railway and subway companies in Tokyo metropolitan areas and (2) those produced jointly by the companies. The total number of posters was 337.

1.4.2 RESULTS AND DISCUSSION

Figure 1.20 below shows the pattern distribution of awareness campaign posters in Japan.

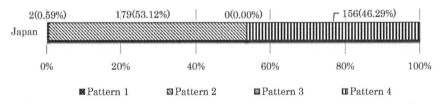

FIGURE 1.20 Pattern distribution in Japan.

First, evidently, Japanese posters are concentrated in Patterns 2 and 4. In other words, Japanese prefer Pattern 2, including the utterances of the discourse level, which is oriented toward the imaginary world level, and Pattern 4, which includes utterances of both the discourse level and the imaginary world level as a dual communication structure, rather than Pattern 1, which mainly uses only descriptions. These characteristics are more obvious in comparison with France in the next section.

Passenger Identity in Public Transport Awareness Campaign Posters   15

The analysis also revealed that Japanese posters frequently depict the characters that appear, not only in one scene, but also in the entire series for one year, and as an image character of a railway company for many years (hereinafter referred to as "a mascot").

As shown in Figure 1.21, a girl named *Miteruchan* (i.e., Miss Watching) circled was used as a mascot 10 times in the 2014 and 2015 series for the Metro Cultural Foundation. As shown in Figure 1.22, the character of the cheering squad with 3 members appeared as a mascot 5 times in the 2014 series (circled). However, as will be discussed later in Section 1.6.2, Japanese posters rarely use realistic images such as photographs, which are often observed in French posters. Figure 1.23 shows the frequency of each character: mascots, realistic images, and descriptions.

FIGURE 1.21
(=Figure 1.8)

FIGURE 1.22
(=Figure 1.4)

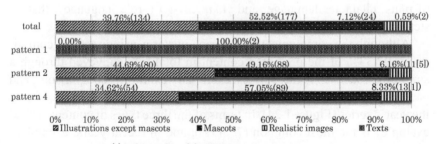

( ) is the number of the posters.
[ ] is the number of the posters with realistic image.

FIGURE 1.23   Distribution of types of images used[3].

---

[3]In the case of Japan, no corresponding examples were found for Patterns 1 and 3; they will be omitted hereafter.

As Figure 1.23 shows, mascots are distributed in more than 50% of the posters in Japan overall, where the mascots commonly use informal speech styles to convey their message in a simple and impactful manner. In Japan, the communication model with these mascots is considered to be effectively used as a strategy to convey messages without damaging the mood of passengers. In Figures 1.21 and 1.22, for example, the mascots appeal to the validity of the message from the mascot's viewpoint as a keyword of "Annoyance to others" to persuade the passengers.

One of the characteristics of Japanese railways is that many private railway companies coexist in a complex manner. The posters related to "manner" among awareness campaigns are generally produced by every company, so the number of posters produced is overwhelmingly large compared to other countries (300 out of 337, 89.02%. See Figure 1.44 below). Furthermore, as they are competitive with each other, one of their features is the use of indirect expressions with the keywords "refrain" and "consideration" for passengers, so as not to damage the image of the company.

As an indirect way of expression, the use of deformed characters is more conspicuous than the use of realistic images. Additionally, the targets displayed include not only characters that require knowledge of their cultural background to understand them, such as those of Japanese folk-tales and of the Edo period, but also characters that can cause topicality, such as world masterpieces, animals, humans characterized as animals, and popular anime characters. The characters sometimes make full use of puns, character language, and *yakuwari-go* ("role language", that is, stereotypical utterances).

Although the communication via each poster was indirect and compli-cated, 97.33% of the presented content in total was conveyed through a simple approach using three styles: exemplary examples, comparisons, and bad examples, based on the analysis of the contents of posters throughout the year's series (Figure 1.24). In terms of the three types of content, a "bad example" is the most common (77.45%), and in recent years, exemplary cases have been increasing gradually. Some companies develop the same number of posters of exemplary examples and bad examples in the series, and compare the two side by side in one poster, but this is still considered as a simple approach.

Passenger Identity in Public Transport Awareness Campaign Posters 17

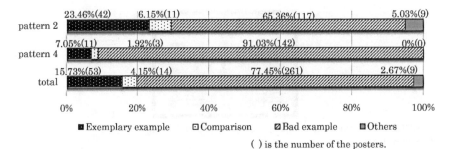

FIGURE 1.24 Distribution of contents presented on posters[4].

For example, in the case of the Metro Cultural Foundation, which produces 1 poster each month, 6 posters each of "exemplary examples" and "bad examples" were produced in 2019 (Figure 1.25). The Tokyo Metropolitan Bureau of Transportation, which produces 5 posters per year, made only comparison posters in 2014 (Figure 1.26).

FIGURE 1.25 Exemplar poster (left) and bad example poster (right)

(*Source*: Reprinted with permission from Metro Cultural Foundation 2019 © 2019, 2020 Metro Cultural Foundation, Tokyo, Japan)
[Left]
S1: Share☺Seki-o yuzurutte, omoiyari-o sheasurukoto-kamo.
"Share☺Giving up your seat may mean sharing your compassion."
[Right]
S1: Wet. Anata-no kasa-de mawari-o nurasanai-yōni shite-ne.
"Wet. Don't get your surroundings wet with your umbrella."

---

[4]In the case of Japan, no corresponding examples were found for Patterns 1 and 3; they will be omitted hereafter.

18   Language and Cross-Cultural Communication in Travel and Tourism

FIGURE 1.26 Poster to compare exemplars and bad examples (=Figure 1.4)
S1: Tenimotsu-wa mawari-eno hairyo-o wasurezuni. (upper side)
"Please be considerate to others when handling baggage."
S2: Suteki-desu! Mawari-o kizukau omoiyari ! (left)
"How nice! When you consider for your surroundings."
S2: Komarimasu! Anata-no nimotsu, shirampuri. (right)
"How annoying! When you pretend ignorance your bag."

Japanese awareness campaign posters tend to indirectly present a message as a whole, but the situation is slightly different when the posters are limited to those that exceed the level of "manner," such as dangerous behavior (Figure 1.27), acts that interfere with operation, and criminal acts (Figure 1.28).

FIGURE 1.27 A poster appealing to people to be careful of accidents on platforms when they are drunk.
(A collaboration by railway operators, 2016–2017 Japan)

FIGURE 1.28 Campaign against sexual harassment on trains.
(A collaboration by railway operators, 2016, Japan)

The awareness campaign posters of this type are generally co-produced by several railway companies, and their number is small among Japanese awareness campaign posters (37 out of 337 posters, 10.97%. See Figure 1.44 below). The co-produced posters of this type tend to present a direct message by: (a) not using puns or deformed characters, (b) concretely showing accident data using %, etc., and (c) depicting specific characters such as the police and crew. However, their composition is the same as that of a normal poster in that the dual structure of Pattern 4 is used and the viewpoints of other passengers and children, who are third parties, are adopted.

FIGURE 1.29

(*Source*: Reprinted with permission from Metro Cultural Foundation, 2017 © 2017 Metro Cultural Foundation. Tokyo, Japan)
S2: Sentō-ga sumaho-ni muchū-de daijūtai. Ushiro-no hito-wa ōmeiwaku-desu-yo.
"A heavy traffic jam occurs due to the person immersed in his smartphone at the front. It's a big nuisance for the passengers behind."

Additionally, in contrast to France, only two Japanese posters were designed from the perspective of a troublemaker (see Section 1.6). Furthermore, as in the thought or free direct speech of Figure 1.29, it is a characteristic of Japanese posters to emphasize that "it is annoying to people" rather than alluding to dangers to the public.

1.5 ANALYSIS OF AWARENESS CAMPAIGN POSTERS IN FRANCE

1.5.1 MATERIALS COLLECTED

In France, unlike Japan, few railroad companies conduct awareness campaigns through posters several times a year, or sometimes monthly. Therefore, the total number of awareness campaign posters found in public transportation in France is not as large as in Japan. For this reason, we collected awareness campaign posters not only from the metropolitan area (Paris and its surrounding areas) but also from 26 cities (30 lines) throughout France. We also expanded the target years to cover the period from 2006 to 2021.

However, the number of posters collected was still less than that in Japan (337 posters), and the total number of French posters to be analyzed was 271. A cultural difference between the two countries can be observed in terms of the number of posters. In other words, Japan tends to offer information and instructions to passengers more frequently than does France.

1.5.2 RESULTS AND DISCUSSION

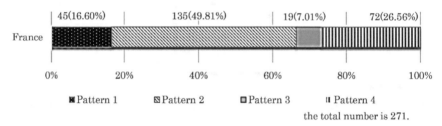

FIGURE 1.30 Pattern distribution in France.

In France, as well as in Japan, Pattern 2 is the most common, accounting for nearly 50% of the total. Adding the number of Pattern 1 to this, we obtain 66.41%, which shows that there is a low tendency to send messages through the utterances of S2, the subject of the imaginary world level.

The dual communication style, which is a complex communication style with both S1 and S2 utterances, accounts for a quarter of the total, while the total number of simple mono communication styles as shown in Patterns 1, 2, and 3 is 73.42%, which shows that a mono communication style is preferred in France.

As for the image contents used in the posters, the distribution of types of images used in French posters is shown in Figure 1.31.

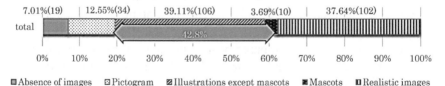

FIGURE 1.31 Distribution of images used.

Passenger Identity in Public Transport Awareness Campaign Posters   21

FIGURE 1.32
(*Source*: Reprinted with permission from Citura, 2018 © 2018 CITURA (TRANSDEV). Reims, France)
S2: A pied, en traversant, je suis vigilant.
"When crossing the track on foot, I am vigilant."
S1: Un cornichon ça fait pourtant toujours attention!
"Even a pickle (=idiot) is always careful!"

The total number of illustrations of irregular characters (39.11%), and illustrations of mascots, that is, standard characters, such as a character circled in Figure 1.32 (3.69%), amounted to 42.8%. However, unlike the Japanese cases, posters with mascots were few (3.69%).

In addition to the illustrations that serve to create a fictional universe, 37.64% of the posters contain realistic images such as photographs of people (Figure 1.33) or anthropomorphic computer-generated graphics. This type of image makes the universe of the posters real and concrete.

FIGURE 1.33
(*Source*: Reprinted with permission from Soléa, 2015 © 2015 Soléa – Realization Mardi stratégie & Créativité. Mulhouse, France)
S2: La fraude, ça craint!
"I'm worried and unsettled about riding without paying."
S1: La fraude c'est des bus et des trams moins souvent!
A partir du 13 avril, contrôles élargis sur toutes les lignes de tram et de bus.
"Fraud means fewer buses and streetcars!
From April 13, extended controls on all streetcar and bus lines."

Finally, there were also cases with pictograms (12.55%) or cases without any image, where only the message was presented without any photograph or character illustration (7.01%). In this context, the poster's universe can be qualified as abstract and universal.

In summary, in France, passengers receive messages in three broad categories of situations: 1) imaginary situations via illustrations (42.8%); 2) real, concrete situations via photographs and computer-generated graphics (37.64%); and 3) abstract, universalized situations via the use of pictograms, or the absence of images (19.56%).

The presentation of realistic and concrete situations is a strategy that can easily evoke familiar experiences in passengers, and make them think that the poster is about them. The use of pictograms is also a strategy to make passengers relate to the contents of the poster, although the approaches are different. The use of pictograms, as opposed to realistic images, eliminates concreteness as much as possible, and does not present a certain imaginary world as character illustrations. Posters that use neutral, highly abstract images such as pictograms are devices that allow everyone to feel involved, without identifying emotionally with either the fictional character illustrated or the particular person pictured.

As a strategy to encourage passengers to self-identify, we also observed in France a case of deploying multiple posters with different pictures of people for a single awareness content.

The series of posters in Figure 1.34 asks people to be careful about the streetcar/bus. Although the message is the same, each poster depicts a different person in a different situation. From left to right: a woman on foot, a young man on a kickboard, a man in a car, and an elderly woman on a bicycle. In France, the emphasis is on making passengers feel involved in the situation.

FIGURE 1.34

(*Source*: Reprinted with permission from Bibus Brest, 2021 © 2021 Bibus Brest, Brest, France)
S1: Attention au tram/bus! Un seul regard suffit pour éviter le pire!
"Beware of the streetcar/bus! One look is enough to avoid the worst!"

Passenger Identity in Public Transport Awareness Campaign Posters 23

Furthermore, concerning the content of posters, a set of posters simply convey information, rather than providing good examples to imitate and/or bad examples to avoid as shown in Figure 1.35. In the case of Pattern 1, this was more than 70%.

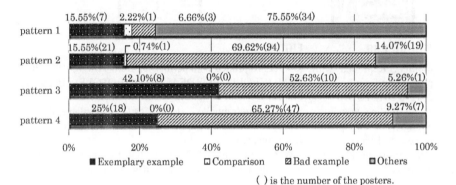

FIGURE 1.35 Distribution of contents exemplified.

The following are examples of each type.

FIGURE 1.36

(*Source*: Reprinted with permission from Tub, 2021 © 2021 Tub – Realization Studio Lannion, Saint-Brieuc, France)
S1: Pliez votre poussette, la place sera plus nette.
"Fold your stroller, the space will be cleaner."

FIGURE 1.37

(*Source*: Reprinted with permission from TaM, 2017 © 2017
TaM – Realization KFH.
Montpellier, France)
S1: Pratique pour les uns, Gênant pour les autres.
"Convenient for some, Annoying for others."
S2: Coucou. Pardon. J'aimerais descendre!
"Cuckoo. I'm sorry. I'd like to alight!"

FIGURE 1.38

(*Source*: Reprinted with permission from Le Met', 2012 © 2012 Le Met' (ex-TCRM). Metz, France)
S1: Dans le bus, pas de place pour l'incivilité!
"In the bus, no place for incivility!"

FIGURE 1.39

(*Source*: Reprinted with permission from TBM, 2019 © 2019 TBM (KEOLIS BORDEAUX METROPOLE), Bordeaux, France)
S1: On vous rappelle… 122€ c'est le prix d'amende pour absence de titre de transport ou non validation du ticket. Je voyage, je valide! (Slogan)
"We remind you … 122€ is the fine for not having or not validating a ticket. I travel, I validate! (Slogan)"

Figure 1.36 shows an exemplary case, and Figure 1.37 shows the case of a bad example. Figure 1.38, however, is not a case of examples, but a statement of the crew's firm stance that they will not tolerate rudeness. Figure 1.39 reminds us that the fine for riding without paying can be as high as €122. These examples do not appeal or suggest that "we should do this because this is what people do (whether by exemplary or bad example)," but rather they state a definite fact to the passengers directly. In France, this strategy is also used to persuade passengers directly.

FIGURE 1.40

(*Source*: Reprinted with permission from Service communication Tub., 2019 © 2019 Service communication Tub. Grenoble, France)
S2: Le chien courait après le chat, et moi je courais après le chien… bref je l'ai atrrapé, et pour venger il a mange mon ticket de bus, … bla bla bla…"
"The dog was running after the cat, and I was running after the dog… in short I caught him, and for revenge he ate my bus ticket, … blah blah blah…"
S1: V'la l'excuse ! La fraude même avec du "bla bla ba…" ça finit toujours par coûter cher !
"Here is the excuse! Fraud even with "blah blah blah…", it always ends up being expensive!"

In this way, France tends to present messages in a concrete situation and in a direct, unmoderated manner, which is also confirmed in posters where the words of passengers or station staff in specific situations are quoted directly. Frequently featured are problematic passenger discourses, as shown in Figure 1.40: of the 271 total cases, 3 cases of words of station staff, 29 cases of words of annoying passengers, 9 cases of words of exemplary passengers, and 10 cases of words of annoyed passengers. At this point, we can see a strategy to make passengers, especially those who cause problems, feel concerned about the cases in question.

Finally, the distribution of actions targeted by awareness campaign posters is as follows: 38.38% of the posters (104 out of 271) considered good manners for a pleasant and comfortable journey, whereas the rest (167 out of 271, 61.62%) were regarding behaviors related to the reader's safety (such as not going onto the tracks), traffic obstruction (such as blocking the door of a departing train), or criminal acts (such as traveling without paying or sexual harassment) (See Figure 1.44 below). We can assume that in France, passengers are considered individuals rather than members who contribute to the building of a cooperative collective society. Thus, passengers are not intervened in their behavior, as long as they do not commit any problematic act that goes against good manners.

1.6 COMPARISON OF THE RESULTS

In Sections 1.4 and 1.5, we analyze the characteristics of awareness campaign posters in Japan and France, respectively. By comparing the results, we can see that the awareness campaign posters in the two countries are designed differently. The main differences are as follows.

1.6.1 DIFFERENCES IN COMMUNICATION MODELS

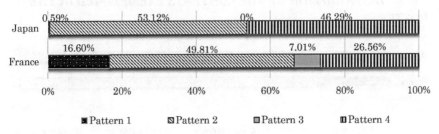

FIGURE 1.41 Pattern distribution in Japan and France.

In Japan, a dual communication style (Pattern 4) is used in nearly half of the posters, whereas in France, a mono communication style—in mainly Patterns 1, 2, and 3—dominates the majority of posters as shown in Figure 1.41. It can be said that the Japanese posters tend to prefer more complex styles of communication.

1.6.2 DIFFERENCES IN THE IMAGES USED

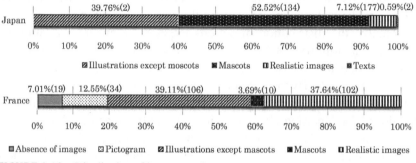

FIGURE 1.42 Distribution of images used.

Posters in Japan make extensive use of illustrations of mascots or characters with low realism, whereas those in France use realistic representations with photos or computer-generated graphics, or consider the universal presentation of situations through the use of pictograms or non-use of images as shown in Figure 1.42. In other words, by using mascots that are less realistic than photographs of passengers as a medium to convey messages both as S2 in the imaginary world and as S1 in the discourse world, Japanese posters are trying to create a distance from the passengers as readers. On this point, the tendency to transmit messages indirectly in Japan can be seen as well.

1.6.3 DISTRIBUTION OF THE CONTENTS EXEMPLIFIED IN THE POSTER

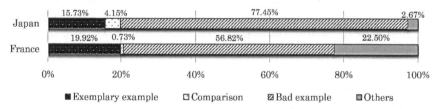

FIGURE 1.43 Distribution of the contents exemplified in the poster.

In both countries, the most common example is the simple presentation of bad examples. In recent years, however, exemplary and bad examples have been used in equal number in the Japanese series of posters, and exemplary and bad examples have been compared in every single poster in the series. This is considered a simple confrontation approach of exemplar versus bad examples.

As a characteristic of France, other examples that do not present exemplary or bad examples were observed in more than 20% of the posters, whereas in Japan, this was found in less than 3%. In France, it can be said that the tendency to present the message in a straightforward manner, without mitigating it by presenting exemplars, is stronger than in Japan. This also highlights the difference between France and Japan, because Japan tries to soften the message and create distance from the passengers as readers.

1.6.4 DIFFERENCES IN THE TARGETED ACTS

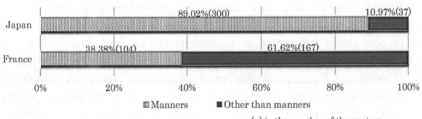

FIGURE 1.44 Manners vs. others.

In Japan, nearly 90% of the awareness campaign posters targeted manners to make people feel comfortable and pleasant among other passengers. In France, on the contrary, the majority of the posters (167 out of 271, 61.62%) were created in relation to the following three points: 1) safety concerns for the passengers, 2) obstruction of train services, and 3) criminal acts (riding without paying and sexual harassment).

Certainly, cultural and social differences contribute to this result, such as the fact that in Japan, there are fewer unpaid rides, and that campaign posters against problematic behavior are created jointly by several companies. However, in France, which directly appeals to every passenger, passengers are treated as individuals who are aware of their responsibilities and, therefore, their behavior is unrestricted unless they go against good manners and behave in a problematic way. In contrast, in Japan, where complex communication styles are used to appeal to users more indirectly, distance from others is emphasized, and many campaign posters convey information about the proper behavior for maintaining appropriate distance from others.

1.7 CONCLUSIONS

In this study, Japanese and French awareness campaign posters were analyzed using a common communication model to show the differences in message transmission strategies between the two countries. The Japanese posters adopted a dual communication style and were characterized by the use of *kawaii* ("cute") and unrealistic mascots to communicate indirectly. However, French posters were based on a mono communication style, featuring more realistic characters, and tended to communicate only to the person concerned in a limited way.

From the above conclusions, clearly, public transportation systems in Japan and France perceive users differently and appeal to them in different ways. In French public transportation, passengers are recognized as individuals who act based on their judgment and responsibility and are directly appealed to. However, in Japan, passengers are recognized as others who should be considered as members of the same *wa* ("harmony") group and the posters indirectly convey messages to them. In this way, passengers who are members of the *wa* group are also required to pay attention to each other.

Although the analysis was based on the communication model, communication on the receiver side was not analyzed in this study, which will be a topic for future research.

KEYWORDS

- **linguistic landscape**
- **public signs**
- **communication model**
- **public transport strategy**
- **cultural differences**

REFERENCES

Adams, J. (1985). *Pragmatics and Fiction.* Amsterdam: John Benjamins Publishing Company.

Banfield, A. (1995). *Phrases sans parole: théorie du récit et du style indirect libre* [Unspeakable sentence: theory of narrative and free indirect style]. Éditions du Seuil.

Bayne, K. (2018). Manner Posters as an Element of the Japanese Linguistic Landscape. *Bulletin of Seisen University Research Institute for Cultural Science, 39,* 61–88.

Choi, J. (2021). Why do anime characters appear and say puns on manner-posters? —An analysis of narrative structure. *JSJS 2021 Conference Handbook.* The Japanese Society for Language Sciences, 69–70.

Fujimura, M. & Taniguchi, A. (2017). densha manā keihatsu posutā-ga manājunshukōdō-ni ataeru eikyō [Influences of awareness campaign posters of manners in trains to good manner movement]. *Dobokugakkai Ronbunshū D3 (dobokukeikakugaku) 73*(5), I_1042. https://doi.org/10.2208/jscejipm.73.I_1033

Mikame, H. (2017). shinteki shitensei to taikenwahō-nitsuite [Mental perspective and free indirect speech]. In T. Hiratsuka (Ed.), *jiyūkansetsuwahō-towa nanika: bungaku to gengogaku no kurosurōdo* [What is free indirect speech: crossroad of literature and linguistics] (pp. 143–192). Hituzi syobo.

Mizuta, Y. (2013). The Generating Mechanism of the Perlocutionary Effects of Train Manner Posters. *Departmental Bulletin Paper of International Christian University*, 55, 149–156.

Wang, J. (2020). Analysis of Catchphrase Composition and Narrative in Movie Posters —Differences between Japanese and Chinese. *Proceedings of the 23rd Conference of the Pragmatics Society of Japan, 16,* 9–16.

Yamamoto, A (2020). Kōkyōkōkoku-no egaku sekai—tēma-no henka oyobi naiyōjō-no tokuchō [A Content Analysis of Japanese public service advertisements]. *Jinbungakubu Kenkyūronshū of Chūbu University*, 44, 47–68.

Yamaoka, M. (2005). *"katari" no kigōron: nichieihikaku monogataribun bunseki* [Semiotics of narratives: a contrastive analysis of Japanese and English fictions]. Enlarged edition. Shōhakusha.

Yamauchi, K. (2004). Kōshūmanā kōjō-no tame-no sokyūhōhō-nikansuru kentō (2) — settokumesuji-no ronkyo-no chigai-ni chakumokushite [Consideration of appeal methods for public manner (2): focusing on differences in arguments]. *Nihonkyōikushinrigakkai Sōkai Happyōronbunshū, 46*(0), 425. https://doi.org/10.20587/pamjaep.46.0_425

Reference Materials

Bibus Brest. (2020). Advertising campaign 2020 *Sécurité routière, redeoublez l'attention* [Road safety, pay attention again]. Brest, France.

Bibus Brest. (2021, November 9). *Un seul regard suffit pour éviter le pire !* [...] [One look is enough to avoid the worst!]. [Image attached] [Status update]. Facebook. https://www.facebook.com/BibusBrest/posts/4793652397345464

Bureau of Transportation. (Accessed November 1, 2021). *Mamorō manā!* [Let's keep our manners together!]. Tokyo Metropolitan Government, Japan.

Posters 2015: https://www.kotsu.metro.tokyo.jp/pickup_information/manner/2015manner.html

Posters 2017: https://www.kotsu.metro.tokyo.jp/pickup_information/manner/2017manner.html

Posters 2020: https://www.kotsu.metro.tokyo.jp/pickup_information/manner/2020manner.html

CITURA (TRANSDEV). (2018, October 22). *UN CORNICHON ÇA FAIT TOUJOURS ATTENTION !* [Even a pickle (=idiot) is always careful! "On foot, when crossing, I am vigilant"]. […] [Image attached] [Status update]. Facebook. https://www.facebook.com/CituraBusTram/posts/1567386073407415

Le Met' (ex-TCRM). (2012). Advertising campaign. Metz, France.

Metro Cultural Foundation. (Accessed November 1, 2021). *Manā posutā* [Manners Posters]. Tokyo, Japan.

Posters 2012: https://www.metrocf.or.jp/jigyou/manner_poster/2012.html

Posters 2014: https://www.metrocf.or.jp/jigyou/manner_poster/2014.html

Posters 2017: https://www.metrocf.or.jp/jigyou/manner_poster/2017.html

Posters 2019: https://www.metrocf.or.jp/jigyou/manner_poster/2019.html

Posters 2020: https://www.metrocf.or.jp/jigyou/manner_poster/2020.html

SEMITAG. (2019). *Rapport d'activité - de M' Tag* [Activity report 2019 from M' Tag] , p.24. [Press release]. https://www.google.com/url?sa=t&rct=j&q=&esrc=s&source=web&cd=&ved=2ah UKEwjYvpCwsKH4AhV--TgGHaCtC0IQFnoECAcQAQ&url=https%3A%2F%2Fs emitag.tag.fr%2Fcms_viewFile.php%3Fidtf=2763%26path=Rapport-Activite-2019. pdf&usg=AOvVaw1D3XUlmLER2-2x9k8ixm32

Service communication Tub. (2019, January 11). Campagne de sensibilisation sur la fraude : « V'la l'excuse ! » [Fraud Awareness Campaign: "V'la l'excuse!"] […] [Image attached] [Status update]. Facebook. https://www.facebook.com/143466229168146/posts/campagne-de-sensibilisation-sur-la-fraude-vla-lexcuse-nouvelle-campagne-de-commu/1105636526284440/

Soléla - Mardi stratégie & créativité. (2015, accessed November 2, 2021). *SOLEA dit stop à la fraude !* [SOLEA says stop fraud!]. https://www.agence-mardi.com/solea-dit-stop-la-fraude/

TaM. [@TaMVoyages]. (2017, March 14). Lutter contre les incivilités. Tous différents ? Soyons bien ensemble ! ☺ Alors respect sur toutes les lignes ! [Fight against incivilities. Everyone is different? Let's be good together ☺ So please respect on all lines! @Montpellier3m #TaMVoyages] [Image attached] [Tweet]. Twitter. https://twitter.com/TaMVoyages/status/841578938870697986/photo/2

TBM – KEOLIS BORDEAUX METROPOLE. [@info_tbm]. (2013, November 12). *Il revient sur le réseau Tbc pour un ultime clin d'oeil #Serge, je monte, je valide* [He returns to the Tbc network for a final wink #Serge, I ride, I validate]. [Image attached] [Tweet]. Twitter. https://twitter.com/tbc/status/400264443411439616

TBM – KEOLIS BORDEAUX METROPOLE. (2018, November 8). *Parce que le harcèlement est un délit et que lutter contre est une priorité, nous pouvons tous agir !* [Because harassment is a crime and fighting it is a priority, we can all take action!]

Passenger Identity in Public Transport Awareness Campaign Posters 31

[...] [Image attached] [Status update]. Facebook. https://www.facebook.com/InfoTBM/posts/10155989955852992/

TBM - KEOLIS BORDEAUX METROPOLE. (2019, accessed November 2, 2021). *Déclaration de performance extra-financière 2019* [Declaration of non-financial performance 2019], p.14. [Press release]. https://www.keolis-bordeaux-metropole.com/sites/default/files/atoms/files/dpef_definitive_2019.pdf

Tisséo. (2020, December 16). *Extrait du registre des délibérations du syndicat mixte des transports en commun de l'agglomeration toulousaine* [Extract from the register of the deliberations of the syndicat mixte des transports en commun de l'agglomeration toulousaine], Séance du 16 décembre 2020 [Meeting of December 16, 2020], p.55. [Press release]. https://www.tisseo-collectivites.fr/sites/default/files/D.2020.12.16.4.1_0.pdf

Tub. (2021, accessed November 2, 2021). *Incivilités dans les bus* [Incivilities on the buses]. https://tub.bzh/infos/actualites/incivilites-dans-les-bus

CHAPTER 2

How Philosophy of Tourism Tussles with Cultural Diversity and Cultural Tolerance

SOORAJ KUMAR MAURYA

Zakir Husain Delhi College (Evening), University of Delhi, New Delhi, India

ABSTRACT

Cultural charms are commonly tied to the host people's distinctive qualities and are frequently associated with a nationwide group or a portion of the populace that has retained their customs. Multicultural or multi-religious nations may have one-of-a-kind features that visitors find intriguing and fascinating. Multiculturalism is a popular tourist destination because it provides people from all origins with an intimate experience. Furthermore, a culturally or ethnically diverse populace can give a linguistically more skilled workforce that is much more attentive to the demands of tourists from other civilizations, resulting in a friendlier place for them. As a consequence of massive international immigration, the problem of a diverse nation has recently been a hot topic. There is a widespread view that cultural or ethnic fractionalization will inevitably result in challenges in communication and cooperation, resulting in inferior economic growth, less sustainable social and economic systems, including, finally, a decrease in productive capacity. The consequent tensions and challenges may scare tourists away, putting the hospitality industry in jeopardy in various cultures. On either side, cultural fractionalization and the consequent cultural diversity might be welcomed as significant assets in the form of

---

Language and Cross-Cultural Communication in Travel and Tourism: Strategic Adaptations.
Soumya Sankar Ghosh, Debanjali Roy, Tanmoy Putatunda, & Nilanjan Ray (Eds.)
© 2025 Apple Academic Press, Inc. Co-published with CRC Press (Taylor & Francis)

34 Language and Cross-Cultural Communication in Travel and Tourism

a diverse body of data, customs, talents, and traditions that can help boost innovation and inspiration. This paper focuses on tourism philosophy, how cultural diversity is related to tourism, and how cultural diversity, including cultural tolerance, substantially impacts tourism. The paper will conclude by advancing the author's position regarding the same.

2.1 INTRODUCTION

Since business travelers, including convention attendees, might mix conferences with tourist-type activities, tourism (along with traveling) is hard to define distinctively; however, generally, a tourist is a momentarily leisured individual who explores a destination distant from residence to witness the transformation of mental or psychical, spiritual, cultural, behavioral, societal, moral, and humane in general (Goulding et al., 2009). According to the Macmillan Dictionary, tourism is a company that offers services to individuals on vacation. According to the World Tourism Organization, individuals traveling to and staying in locations outside their typical surroundings for no more than one year for enjoyment, business, or other objectives, in terms that go beyond the popularly held idea that tourism is restricted to holiday activities alone.

Moreover, culture is very abstract, unlike civilization, and consequently hard to define; however, in a very simplistic way, according to the "United Nations Educational, Scholarly, and Cultural Institution, culture is the series of unique divine, content, intelligence, as well as sentimental characteristics of society or a social group, including, in addition to literature and art, preferences, aspects of dwelling together, value- systems, virtue-system, traditions, and beliefs, and it includes, furthermore to literature and art, behaviors, ways of living around each other, moral codes, customs, and beliefs." Concerning tourist philosophy, "Society" can be analyzed as a group's system of social interactions, while "culture" is described as the firm's shared beliefs and emblems. Culture is the accumulation of information, views, experience, values, meanings, attitudes, hierarchies, perceptions of time, religion, geographical linkages, roles, and universe ideas. According to Mathew Arnold, culture is "acquainting oneself with the greatest that has been understood and uttered in the world by ourselves."

In overview, the way of life as folks understand its situation is the consequence of a continuing dialectic among a variety of stakeholder speakers, such as but not restricted to political groups, religious organizations, economic interests, world media, ruling classes, academic facilities, as well as, of course, the tourism sector and visitors, as well as the everyday activities of individuals going about their daily lives. This culture-forming process applies to our perceptions of the past and our handling of the environment and modern civilizations' cultures.

Moreover, tourism is both a social and cultural phenomenon. It influences cultures and civilizations and is formed by them; that is why it comes under the investigation of philosophy. When it comes to the link between tourism and philosophy, it is frequently challenging to underline the difference between cause and effect. This is exacerbated further by the reality that philosophical understanding to explore the world gives new wings to understand and analyze the environment out there and enables to explore a new flight in the world of speculation and reflection of the tourism. It does change over time, notwithstanding environmental activists' and environmentalists' best attempts to museumize monuments and landscapes to preserve their originality.

Philosophical vision also undertakes questions such as, what impact does tourism have on the surrounding area? Tourists and tourism have a significant effect on the host nation. Moreover, hospitality is a decisive and unique factor for community improvement (Surugiu and Surugiu, 2013). The most important influence that tourism may have on the private home, domestic life, emotional states, firmly held values, and the moral fiber of the collection of individuals who make up the society. The effect of social and cultural concepts on the younger crowd who have had intimate interaction with foreign visitors is increasingly noticeable.

Furthermore, culture and tourism have a usually positive relationship that may increase the attraction and profitability of a location, region, or country. Culture is becoming a crucial fundamental component of the tourist sector in a crowded global economy. On the other hand, tourism is an essential instrument for enriching the culture and producing cash that may be utilized to promote and develop culture, history, manufacturing, and innovation. Due to their evident synergies and development possibilities, culture and tourism are intertwined. Cultural tourism is one of the fastest expanding segments of the global tourist industry, and the recreation and tourism sectors are rapidly being leveraged to advertise places.

2.2 PHILOSOPHY OF TOURISM

Philosophy is the study of people's actions, experiences, and relationships. The subject advocates that a person is always impacted by a group of individuals, while ecological philosophy is concerned with the effects of the physical environment on human behavior. Social, environmental, and philosophical discoveries are applied in tourism better to comprehend travelers' behavior, emotions, and motives. Not just anyone, but the increased focus on the "wide theme of consciousness, which relates to human cognition and data processing, which provides crucial relevant concepts for tourist study and analysis," To preserve visitors' health and well-being, it is vital to examine tourism not only from an economic-geographical standpoint but also from the perspective of their behavior when confronted with emotional issues (Pascariu and Frunză, 2012). Tourists often seek new perspectives of life, new experiences, and originality or unexpected circumstances to avoid their daily routines and stress. According to Virdi and Traini (1990), every tourist's behavior represents personal and social elements of his leisure time and incentive.

Furthermore, every such action entails not just a financial but also a human emotional commitment. Given that a tourist occupies his leisure time in a setting that he has selected, and that is distinct from what he is accustomed to in daily life, tourism philosophy defines a tourist's character based on an examination of his behavior and activities in such an "alternative" setting. Tourist behavior is influenced by emotional, social, motivational, and cognitive factors, allowing for extensive psychological surveillance. Desire (needs), anticipation, decision-making, fulfillment, expertise, and relationship analysis are essential aspects of tourist philosophy.

Tourist philosophy makes a humble attempt to understand human requirements regarding the encounter. A traveler or tourist comes across from a specific journey on one side, and a real connotation regarding tourist services on another is critical. It is not only about motivation; it is about the interconnectedness of individual characteristics, where contentment and discontent retroactively construct the mental image of a goal, which drives the client's requirements and influences his future motivation and behavior. Maslow's hierarchy of requirements, which proposes that individuals are driven to meet fundamental wants before advancing to more sophisticated demands, may also be used in tourism. But how do the many levels of Maslow's hierarchy of requirements connect to the fundamental principles of tourist cognition?

2.2.1 PHYSIOLOGICAL AND SAFETY REQUIREMENTS ARE MET

Each tourist site is universally recognized as having to address two essential criteria: physiological and safety and protection. Physiological requirements: Gastronomy—a wide range of flavors and a high degree of quality that corresponds to the local cuisine. Provincial branding is often added to items to emphasize the product's as well as place's distinctiveness. Quality lodging ensures that visitors' health is not jeopardized. Safety requirements: this reflects the nature of the place, its social activities, as well as the possibility of undesirable pathological characteristics (Pereira Roders and Von Oers, 2011). Different levels of Maslow's hierarchy of wants vary depending on the circumstances in which a visitor may emerge. Culture (tourists seeking local history, customs, or art), active engagement (sports activities), adventure (today referred to as adrenaline), conformism (searching for actions he is familiar with), rest and wellness (wellness), or social standing are examples of these.

2.2.2 FULFILLMENT OF SOCIAL REQUIREMENTS

The very essence of human existence is to accommodate societal experiences, which leads to particular demands—moreover, the social needs, such as a desire to be a group member. Tourism philosophy and psychology maintain track of hospitality, effectively promoting itself without appealing, and how to handle (or rather prevent) any problems between individuals or groups whenever a local region is interested in tourism. It might also be a feeling of belonging to a particular neighborhood and local customs, habits, or connections. It is a fleeting sensation of belonging to a group, such as being a member of a tourist family in the case of visitors. The core premise of tourism, and frequently the primary reason for travel, is to meet people from different countries, regions, or cultures and share personal experiences, attitudes, or opinions. On either side, such well-planned excursions with meticulously planned itineraries may result in unwanted tension, leading to antisocial behavior. Negative behavior may also occur when people are compelled to assist others, as in group outings.

2.2.3 SELF-REALIZATION AND FULFILLMENT OF SELF-APPRECIATION REQUIREMENTS

This level examines customer contentment connected to Maslow's pyramid's 4th and 5th levels. On the one extreme, the relationship between a tourist service company and a customer is remarkable. Even though the service's focus is on the consumer, not the product he or she is using. As a result, a customer's contentment needs loyalty to specific goods, services, locations, or service providers. These amenities pave a better way to explore the place to the desirable extent and fulfill his or her desires and objectives, resulting in the sense of self-realization and self-actualization (Pereira Roders and Von Oers, 2011). This degree of psychical contentment eventually leads to a positive one-ness of the individual toward the environment. On the other hand, dissatisfaction with the goods or services consumed produces internal personal conflicts. The location, organization, and quality of tourist services provided determine the degree of self-realization. It is possible that having too much powerful experience is hazardous. "Is virtually everything all right?" we could recall hearing. It may have an unfavorable effect. Consequently, psychological and philosophical aspects should be included in the interaction between the tourist and traveler's environment of the place that is most tailored to the client's needs and desires, all within the context of his "perfect holidays" notion.

2.3 TOURISM AND CULTURAL DIVERSITY

2.3.1 DELINQUENT FRAMEWORK

There has been growing worry that immigrants from cultural origins substantially dissimilar from those of the host nation may create severe issues, resulting in conflicts, and that integrating immigrants from various backgrounds may become impossible. As per popular belief, if immigrants possess religious, ethnic, and cultural backgrounds comparable to those of the host nation, blending will be more straightforward, but the barrier will be too difficult to overcome. For instance, a new, ethnically diverse immigrant populace refuses or cannot integrate by adhering to their traditional home culture and traditions and getting married to their subgroup. In that case, ethnic distinctions may persist long after the immigrants have arrived in the host nation, and cultural diversity is retained at all stages of its life cycle. Cultural or ethnic fractionalization may lead to problems with communication and collaboration, ethnic disputes, poor economic efficiency, a less stable social and economic environment, and, eventually, a drop in economic production. Ethnic and cultural variety, on either extreme, may be helpful via the diversity of information, customs, abilities, and practices, which can lead to new ideas and innovation (Chang, 2010). Tourism, being one of the country's most significant industries, is especially sensitive to variety. Visits to friends and family, healthcare, leisure and recreation, and an appreciation of nature and history are all examples of tourist spots. As a significant tourist motivator, cultural tourism involves trips to both physical and intangible cultural assets to gather new knowledge and experiences to meet the cultural demands of tourists. Culture and traditions are often tied to the host community's distinctive characteristics and are associated with a particular ethnic group or a portion of the population that preserves its customs. Multiculturalism is a popular tourist draw because it provides guests from all walks of life with a genuine experience. Furthermore, an ethnically or linguistically diverse population may give a linguistically more skilled labor force that is more attentive to the demands of tourists from other cultures, resulting in a more pleasant atmosphere for them. The findings of a quantitative examination of 129 nations that analyzes several diversity measures and examines their effects on the Travel and Tourism Competitiveness Index are presented in this article.

2.3.2 DIVERSITY AND ECONOMIC GROWTH

For decades, the influence of human resources on a country's degree of advancement, including financial improvement, has been a hot area of inquiry. Theoretical considerations show that the population's diverse ethnic makeup may be a beneficial human resource, with improved creative capability, skill, and creativity improving the economic performance of organizations, businesses, and hence regions and nations. Social diversity may sometimes have a detrimental economic consequence. It can impact individual choices since people tend to place more value on events that benefit their social groupings. Personal tactics are also affected since people want to work in homogenous surroundings to save transaction fees. Ethnic variety may influence the production process because although variability and various pools of talents and abilities may boost output, a lack of comprehension and unwillingness to collaborate may reduce it. Education may amplify or decrease the positive or negative economic effects of diversity. Education may contribute to more tolerance and coping with variety and increase or decrease discrimination and prejudice. Religious and ethnic acceptance are essential societal ideals that are often linked to cultural asset conservation.

2.3.3   EFFECTS OF FRACTIONALIZATION ON FINANCIAL PRESENTATION

The majority of study findings seem to concur that increased variety, particularly ethnic diversity, frequently correlates to poorer GDP per capita growth. In 2003, the Human Development Index (HDI) was shown to associate with societal heterogeneity negatively, but not ethnolinguistic heterogeneity, but religious fractionalization, although other research indicated beneficial effects of religious variety on HDI in 2014. Given that one benefit of increased diversity may be a wider pool of creative resources, it is acceptable to investigate if it impacts a country's innovation or competitiveness. According to Ogden et al. (2014), staff cultural diversity is positively associated with product creativity. Hlepas (2013) looked at how cultural and ethnic variety affected the Global Competitiveness Index (GCI), the Human Development Index, and people's confidence in others (Eeckels et al., 2012). Primarily qualitative study revealed no straightforward trends that many industrialized nations are often diverse and that undeveloped countries are often homogenous. He claims that GCI is a stronger indication of economic success than GDP/capita since GDP/capita measures just production and consumption. Still, GCI may show other significant components and capabilities of a nation underlie high or low gross domestic product values. DiRienzo investigated the effects of fractionalization on the Global Competitiveness Index and discovered that racial fractionalization had a negative impact, linguistic fractionalization had a beneficial impact, and spiritual fractionalization had no effect. Comparative assessments in 2007 found positive results of diversification of overseas tourists after controlling for economic liberty, democracy, and gross domestic product. On the other hand, others observed in 2014 that ethnocultural and language fractionalization had adverse effects on tourist competitiveness, but religious diversity had no influence. The contradictory findings suggest that variation in the year between fractionalization information and the response variable (indicating a potential temporal lag) and variations in fractionalization degree from year to year may influence fractionalization's effects in the year between the fractionalization information and the response variable (indicating a potential temporal lag).

2.3.4 TOURISM AND SOCIOCULTURAL DIVERSITY: ADVANTAGES OR DRAWBACKS

In the realm of tourism, ethnic and cultural tourism has been on the rise. Cultural tourism's core incentives are built, physical items and ideals exhibited in daily life, including events and festivals. Old, archeological, architectural, and religious sites, such as countryside buildings, battlegrounds, and historic cemeteries, contribute to heritage tourism. Ethnic tourism may be divided into two categories: one is driven by a sense of homesickness and nostalgia for one's birthplace, and the other is motivated by a desire to discover one's ancestry. The alternative is to visit a community of a minority group and experience a genuine culture, either in faraway regions or inside one's nation, to learn about and comprehend a weird, unique, exotic culture. As observed in Chinatowns, Little Tokyo, and Thai Towns throughout the globe, culture may be a tourist draw in, including high culture, various forms, and, increasingly, popular culture and ethnic culture.

Likewise, Harlem shows that tourism-based growth may be a beneficial motivator for financially disadvantaged neighborhoods. It began to develop as a tourist attraction in the 1980s, depending on the cultural, musical, and recreational heritage of Black America. Even though the additional number of tourists did not initially correlate with improved tourism expenditure, diversity has become a financially appealing choice due to the congestion of conventional tourism sectors. The tourism industry has evolved beyond museum trips, theater visits, and high-art activities to encompass anthropological concerns, having ethnicity and culture at the forefront of this perspective (Pascariu and Frunză, 2012). The 2001 "I Love New York" ad illustrates this trend with its concept of culture and different linkages to European, African-American, Hispanic, Asian, and Native-American cultures. In the industrialized world and Asia-Pacific, ethnic minority culture—comprising tangible and performance culture and customs—has become a prominent tourism focus. The international society of consumption and production has resulted in cultural homogenization, necessitating a need to seek out "differentness" via tourism deliberately. Ethnic characteristics may be seen as a valuable resource, and tourism can help to encourage their rehabilitation, conservation, and fictitious re-enactment.

Multicultural tourism and historical heritage are closely related, although ethnic tourism emphasizes contrasts between the host and visitor communities, while heritage tourism emphasizes one's history. Tourists generally see ethnic minorities as backward and primitive, as shown by their visits to ethnic theme parks. The theoretical underpinnings of cultural tourism are discussed, with some excellent examples from the Far East to exemplify them. Ethnic variety also adds to the diversity of spiritual buildings—churches, mosques, and shrines. Even though Sydney is frequently referred to be "the globe in one town," the metropolis' tourist advertising has seldom emphasized its variety. Immigrant populations have often contributed to a host society's cultural, intellectual, and economic security. Immigrants in transnational communities generated through immigration are typically considered passive receivers of social assistance and active agents of economic regeneration, as shown by instances in the United Kingdom and Canada (Chang, 2010). The popularity of Asian restaurants and cultural events celebrating Indian, Pakistani, Caribbean, and other Afro-American cultures' cuisine, beverages, music, and craftwork demonstrate this. Even though cultural variety as a tourist spot is well accepted, there has been little research on tourism and variety connections. For instance, Das and DiRienzo (2009) looked at the association between the Travel and Tourism Competitiveness Report and ethnic fractionalization.

Whenever the link was adjusted for financial freedom and democracy, they discovered a significant unfavorable association between cultural minorities from 1985 to 2001 and the logarithm of the TTCI for 2009. Higher salaries, on the other hand, may offset the detrimental effect of increased cultural minorities on tourist competitiveness, according to the research. Tourism competitiveness is linked to the integrity of the development of the surroundings, transportation, basic facilities, health care establishments, and a variety of other variables that need national expenditures. Such investments are inevitably dependent on the amount of national revenue on the one side and societal acceptance for such large-scale investments on the other (Ekanayake and Long, 2012). This help is increasingly harder to get by in a more fragmented society. Bacsi (2017c) found similar results using GDP and TTCI data sets from 2014 to 2017. The religious background of a country has a role in worldwide tourist destination selection. The spiritual resemblance has solid predictive strength in the tourism industry. The existence of a shared religious minority in the originating and destination nations has a favorable influence on tourist patterns, according to a cross-country research of 164 countries conducted from 1995 to 2010.

2.4 TOURISM AND CULTURAL TOLERANCE

Danella and Dennis Meadows of Massachusetts produced the paper "Constraints of Expansion" in the year 1972, in which they looked at the impact of growth in the economy on the globe and its development. They substantiated their assertion that the world would fail to deal with the alarming pace of resource usage and environmental damage caused by present economic expansion using computer simulations. According to the systems perspective, the business system requires immediate and radical modifications to achieve a global equilibrium condition.

Furthermore, the scholars favor the view that our planet's existence is a subject of significant worry due to rising overcrowding. This scenario is far from sustainable despite the scarce environmental supplies accessible. One of the earliest publications on the notion of sustainable development was the "World Conversation Strategy," published in 1980 by the "International Council for the Discussion of Nature and Natural Resources." The official "Report of the World Commission on Environment and Development: Our Common Future," often known as Brundtland's report, was issued in 1987 by the International Committee for Environment and Sustainable Development (Caglayan et al., 2012). The article stresses the importance of sustainable development, describing it as redemption. That report was founded on her well-thought-out theory that humans did not rule the Planet from their forefathers but were instead required to pass it on to the next generations.

There has been an increasing international awareness of sustainable development since 1987. The 1992 Rio de Janeiro Summit and "Agenda

44 Language and Cross-Cultural Communication in Travel and Tourism

21" and environmental heating concerns like "global warming" and "smog," which significantly affected Southeast Asia in 1997, showed great motivations to continue the process (Dwyer et al., 2010). At the same time, those years witnessed a tremendous inflow of visitors—a reality that had little to no positive impact on the host nation's local ecosystem.

Contemporary researchers willingly adopted the principles of sustainable development to the tourist industry after they were translated into many realms of human action. J. Swarbrooke was the first to stress the influence of responsible ecotourism, social and natural elements of an area, and inhabitant behavior in a location in 1999. Since then, much work has been done in integrating sustainability themes into tourism studies. Sustainable tourism necessitates a shift in focus from technical difficulties to societal and ethical concerns. The requirement of devising effective techniques of the conceptual and applied possible solution—potential in the concept of lasting peace variety of social societies, social groups, as well as specific persons in an incorporated cultural environment—leads to an ideology of acceptance in the way of solving xenophobia toward "the Other."

2.5 CONCLUSION

Philosophy (*bhāratīya darśan*) observes everything in totality to have a holistic vision and understanding of life, which can be substantiated in tourism philosophy. For instance, society and societal superstructure and base structure are full of transitoriness, and changeableness, which advocates that variety, diversity, and multiplicity are the factual truths of the society and societal experiences and vision, yet and often seen as an issue, a danger, or a source of contention. However, from the above-mentioned approach, the paper has undertaken that ethnic and linguistic heterogeneity does not significantly influence tourist competitiveness, implying that ethnolinguistic disparities do not threaten the tourism industry's success. On the other side, spiritual multiplicity and diversity have been proved to be advantageous when it comes to international experiences of life and the universe. A more diversified populace may be more welcoming to international tourists. Religious and spiritual diversity, variety, and multiplicity may be beneficial since it fosters acceptance of diversity within or outside the geographical boundaries (Kim et al., 2011). Religious and spiritual legacy may be tourist destinations in the type of religious structures, religious locations, and religious holidays. For instance, India has been a torch-bearer when it comes to spirituality and religious diversity that has attracted tourists and travelers since time immemorial. With more variety, the range of this kind of attraction may be more affluent. Ethnicity or dialect did not affect the results. However, they may also bring tourist charms, language and comprehension issues may negate the benefits of variety. It is also worth analyzing if more broad measures like the nation's regulatory arrangements or educational system are appropriate for its real social variety.

The same institutional framework may be perfectly acceptable for a relatively homogenous country in a more diversified nation depending upon the philosophy of the culture, such as how much tolerant the culture is for others (Pascariu and Frunză, 2012). Still, it may be a severe impediment to commercial activity for particular sectors of society in a more diverse nation. As the negative effect of the year of freedom demonstrates, countries that have recently gained independence may have less efficient institutional frameworks to deal with difficulties posed by numerous social groupings. Increased ethnic pride and sensitivity to national identity may result from the shorter history of independence, which may easily lead to disputes or antagonism. These findings are incongruent, and they change depending on the year of the data used to compute religious fractionalization indices. The paper has also included the population effect, although it had no significant influence. As a result, a giant population does not always imply a more creative work-force. A nation's economic climate and institutional framework play a critical part in ensuring that all citizens, irrespective of race, cultural, or religious affiliation (Surugiu and Surugiu, 2013), have accessibility possibilities. The competitiveness of tourism seems to be unaffected by linguistic or ethnic fractionalization. This begs the question of how ethnic variety is exposed: it might be conveyed via language, which is a commodity, as long as everyone speaks the dominant language. Separating ethnic identification from language, on the other hand, may operate as a differentiating factor, leading to increased nationalistic behavior or even hostility.

KEYWORDS

- tourism
- cultural diversity
- cultural tolerance
- multiculturism
- ethnicity
- race
- religion

REFERENCES

Goulding, P.; Theodoraki, E.; Thomas, H. Book Reviews, *J. Policy Res. Tour. Leisure Events* **2009,** *1* (3), 281–285.

Surugiu, C.; Surugiu, M. R. Is the Tourism Sector Supportive of Economic Growth? Empirical Evidence on Romanian Tourism. *Tour. Econ.* **2013,** *19* (1), 115–132.

Pascariu, G. C.; Frunză, R. Corporate Social Responsibility and Sustainable Development of Tourism Destinations—An Analysis from the Perspective of the Developing Regions in the European Context, *CES Working Papers* **2012,** *4* (4), 772–794.

Pereira Roders, A.; Von Oers, R. Editorial: Initiating Cultural Heritage Research to Increase Europe's Competitiveness. *J. Cult. Herit. Manag. Sustain. Dev.* **2011,** *1* (2), 84–95.

Kim, A. K.; Airey, D.; Szivas, E. The Multiple Assessment of Interpretation Effectiveness: Promotig Visitors Environmental Attitudes and Behaviour. *J. Travel Res.* **2011,** *50* (3), 321–334.

Chang, L. C. The Effects of Moral Emotions and Justifications on Visitors Intention To Pick Flowers in a Forest Recreation Area in Taiwan. *J. Sustain. Tour.* **2010,** *18* (1), 137–150.

Eeckels, B.; Filis, G.; Leon, C. Tourism Income and Economic Growth in Greece Empirical Evidence from their Cyclical Components. *Tour. Econ.* **2012,** *18* (4), 817–834.

Pascariu, G. C.; Frunză, R. Corporate Social Responsibility and Sustainable Development of Tourism Destinations—An Analysis from the Perspective of the Developing Regions in the European context. *CES Working Papers* **2012,** *4* (4), 772–794.

Chang, L. C. The Effects of Moral Emotions and Justifications on Visitors Intention to Pick Flowers in a Forest Recreation Area in Taiwan. *J. Sustain. Tour.* **2010,** *18* (1), 137–150.

Ekanayake, E. M.; Long, A. E. Tourism Development and Economic Growth in Developing Countries. *Int. J. Bus. Finance Res.* **2012,** *6* (1), 51–63.

Caglayan, E.; Sak, N.; Karymshakov, K. Relationship between Tourism and Economic Growth: a Panel Granger Causality Approach. *Asian Econ. Financ. Rev.* **2012,** *2* (5), 591–602.

Dwyer, L.; Forsyth, P.; Spurr, R.; Hoque, S. Estimating the Carbon Footprint of Australian Tourism. *J. Sustain. Tour.* **2010,** *18* (3), 355–376.

Kim, A. K.; Airey, D.; Szivas, E. The Multiple Assessment of Interpretation Effectiveness: Promotig Visitors Environmental Attitudes and Behaviour. *J. Travel Res.* **2011,** *50* (3), 321–334.

Pascariu, G. C.; Frunză, R. Corporate Social Responsibility and Sustainable Development of Tourism Destinations—An Analysis from The Perspective of The Developing Regions in the European context. *CES Working Papers* **2012,** *4* (4), 772–794.

Surugiu, C.; Surugiu, M. R. Is the Tourism Sector Supportive of Economic Growth? Empirical Evidence on Romanian Tourism. *Tour. Econ.* **2013,** *19* (1), 115–132.

Teh, L.; Cabanban, A. S. Planning for Sustainable Tourism in Southern Pulau Banggi: An Assessment of Biophysical Conditions and their Implications for Future Tourism Development. *J. Environ. Manag.* **2007,** *85,* 999–1008.

CHAPTER 3

Intercultural Communication in Tourism During COVID-19 Pandemic: Analyzing Audience Appeal of Indian Tourism Campaign Through Social Media

RUMA SAHA[1] and LAKHAN RAGHUVANSHI[2]

[1]*Department of Journalism & Mass Communication, Manipal University Jaipur, Rajasthan, India*

[2]*Department of Journalism & Mass Communication, Devi Ahilya Vishwavidyalaya, Indore, India*

ABSTRACT

Intercultural communication studies how different culture affects communication between two different people of different social group or cultural group. Tourism has always promoted intercultural communication among travelers since ancient times. There is evidence of travelers of ancient times who traveled to different countries whose culture, language, religion, and beliefs were very different from their land of origin, and they not only did intercultural communication but documented it in book form. For example, ancient travelers like Hien Tsang and Fa-Hien who came to India during the ancient period and has documented their experience of intercultural communication in the form of a book which was later used by historians to know various facts about that period. So, traveling always involves some intercultural communication. Even today, though we are all global citizens

---

Language and Cross-Cultural Communication in Travel and Tourism: Strategic Adaptations.
Soumya Sankar Ghosh, Debanjali Roy, Tanmoy Putatunda, & Nilanjan Ray (Eds.)
© 2025 Apple Academic Press, Inc. Co-published with CRC Press (Taylor & Francis)

and the world is a global village traveling is associated with intercultural communication. This paper aims to study and analyze the audience appeal of five tourism campaigns on the social networking site Facebook during the recent COVID-19 pandemic period of 2020–2021. The study will explore how audiences find it appealing in any tourism campaign on social media and virtually feel connected to different cultures and allure to make travel destinations. Given the current pandemic situation when there is lots of bars on international travel destination Indian travelers prefer to explore destinations within the country. Throughout the pandemic, social media and social networking sites played an important role in advertising campaigns in every sector same goes for the tourism sector. Many travel bloggers use travel videos and promote the place apart from traditional tourism advertising campaigns by Incredible India. Our study will explore this area by studying various travel bloggers on Facebook and how they use posts, images, and videos to appeal to travelers to travel to the destination. The methodology applied will be qualitative where social media content will be analyzed and a case study will be done for four such travel campaigns to explore how intercultural communication is a big appealing factor among travelers. The period of study is 2020–2021 during the COVID-19 pandemic. The result will help the travel industry to focus on various appeal factors that influence the audience when they come across a tourism campaign. It will also explore if travelers in the recent time of pandemic are seeking intercultural communication as a motivating factor to travel to different destinations. In the age of social media and social networking sites, most people are hooked online so this study will reveal how intercultural communication takes place in that sphere concerning tourism campaigns.

3.1 INTRODUCTION

The travel and tourism sector is a significant contributor to the growing GDP of the world economy. This sector has generated many jobs and has been consistently growing with improved technological and network integration to the service. World Travel and Tourism Council has noted that the tourism sector has grown by 3.5%, contributing 8.9 trillion USD to world GDP, and has also generated 330 million jobs worldwide by 2019. Similarly, in India the tourism sector growth has been consistent till 2019, contributing to approximately 87 million jobs generation in the domestic market according to the FICCI report of June 2020. An added advantage of Indian tourism is the diverse landscape that helped in offering multiple tourism offerings ranging from luxury travel to adventure, wellness travel and several more. In 2019, India recorded 10.8 million tourist arrival, which was 3.2% higher than in 2018. Due to the pandemic, this industry, along with other sectors, came under severe economic pressure because of lockdowns and restrictions. According to the World Travel and Tourism Council report of June 2020, 2.7 trillion USD loss was incurred by the travel and tourism industry due to the pandemic and job loss was 100.8 million throughout the world (FICCI Report, 2020).

The government of several countries including India has taken several measures like short-time relief measures, working in close collaboration internationally with public and private parties to limit the impact on the tourism industry, which is marked by rapid joblessness (FICCI Report, 2020). The pandemic has disrupted international travel and the supply chain in an unprecedented manner. Many countries banned international travel and did not issue visas even for business travel, thereby bringing the travel sector to a screeching halt. It affected the overall economic shrinkage as predicted by IMF.

In India, COVID-caused lockdowns started at the end of March 2020. The restrictions on public movements compelled the travel and tour sector to shut down completely. According to the FICCI report of June 2020, the loss figure in India is estimated at 16.7 billion USD in the travel and tourism sector. This sector is the largest producer of jobs in the country, which is on the verge of layoffs, and almost 40–50 million that which are directly or indirectly related to the sector was at risk in 2020. The Ministry of Tourism of the Government of India created a "National Tourism Taskforce" to tackle this situation. According to the Centre of the Aviation, the aviation sector in India was yet another sector that incurred huge losses, amounting to 3.6 billion USD between April 2020 and June 2020. A 47% decline in passenger traffic was recorded due to the pandemic. This resulted in flight cancellations indefinitely, later becoming the reason for lay-off of Air staffs by the airlines companies. According to the FICCI report of June 2020, at this juncture, a combination of financial, fiscal, and monetary measures can save the business and people associated with the industry to survive. Therefore, to be able to design the correct policy of action government needs to understand and study the problem carefully.

Amid all the regulatory and fiscal policies that were suggested in the FICCI report of June 2020, it had requested the Government of India to promote a "niche travel product" under the banner of the Incredible India 2.0 campaign. It suggests ted promotion of wellness tourism, religious circuit, culinary tourism, self-drive holidays, caravan tourism, film city tour, and band each tourism. Aggressive marketing across platforms was suggested in the report, with special reference to improving digital content related to travel destinations. There is a lack of adequate digital content related to travel destinations, specifically for domestic locations. The report suggested increasing the flow of digital content, which can be used by travel agencies for promotion and marketing. Increasing visibility can bring more consumers. Additionally, providing a facility for immediate online booking of the destinations through the integration of sites promoting it and booking it can bring positive results.

The objective of this study is to find how digital content promotion with special reference to intercultural communication has been implemented for the domestic tourism market through social media. Moreover, travel bloggers and influencers play an important role in this promotional task. Intercultural communication is another aspect of niche tourism, which attracts tourists to visit India as India is a country of diverse cultures. During a pandemic, several tourism campaigns were undertaken to revive the Indian domestic travel and tourism market. This paper undertakes a case study of such hashtags and campaigns on social media launched during the pandemic period in India from April 2020 to May 2021 by Indian brands as a part of social media promotion to reach out to diverse audiences. The study also includes an analysis of the official Facebook page of two selected Indian travel bloggers and their promotion techniques, during the lockdown period in India between May 2020 and August 2020.

3.2 EXPERIMENTAL METHODS AND MATERIALS

3.2.1 REVIEW OF LITERATURE

Research has been carried out on "Intercultural communication in tourism" by Cristina E. Albu in 2015. The researcher focused on the understanding meaning of intercultural communication in the context of the tourism sector. The paper discusses how this experience of intercultural communication helps the tourist to understand not only another culture and adapt to people of that culture but also understand their own culture in a better way (Albu, 2015).

A recent research paper, titled "Gastro-diplomacy in Tourism: Capturing hearts and minds through stomachs," explained the term "gastro diplomacy" and how food is used as a tool to attain the goal of cultural diplomacy in tourism (Nair, 2021). Gastro-diplomacy is a part of soft power that supports the establishment and promotion of ethnic restaurants, and food festivals to enhance the international presence of diasporas. Further, the research tries to examine the role of national tourism campaigns in the promotion of gastro-diplomacy may in tourism in India. The findings revealed that a government-led tourism campaign is capable enough to introduce places, locations, and specific cuisines to foreign tourists. The researcher has also identified the features of national tourism campaigns that may help in enhancing the gastro diplomacy of India in the future (Nair, 2021).

Gulati's research paper on "Social and sustainable: exploring social media use for promoting sustainable behavior and demand amongst Indian tourists" in 2021 reveals that there is a knowledge gap between tourists and researchers on the topic of "sustainable tourism consumption." The study brings into light the gap and has explained conceptual frameworks for empirical testing of how social media can create and promote tourism demand sustainably. Findings revealed variance in travel habits is not significant with social media usage. Further, it has been found that promotion through social media creates awareness and enhances sustainable tourism. Therefore, social media promotion creates sustainable demand among tourists (Gulati, 2021).

Thomas et al. (2021) researched "Examining the effect of message characteristics, popularity, engagement, and message appeals: evidence from Facebook corporate pages of tour organization signs." Researchers investigated the use of Facebook in tourism promotion by national and state organizations. They analyzed the active Facebook pages page of five tourism departments and have done a content analysis of the posts on those pages. Findings revealed positive emotional appeal that is quite effective while dealing with the experiential tourism products. Moreover, the acquisition of a higher figure of social media followers does not enhance better post engagement (Thomas et al. 2021).

3.2.2 RESEARCH GAP

There is research done in the area of intercultural communication and tourism, gastro-tourism from an Indian perspective, sustainable tourism using social media promotion, and tourism campaigns through Facebook corporate pages by Indian organizations. But there is hardly any research in the area of intercultural communication in the context of tourism during the COVID-19 pandemic concerning Indian tourism campaign promotion through social media.

3.3 THEORETICAL FRAMEWORK

3.3.1 COMMUNICATION ACCOMMODATION THEORY

Howard Giles propounded the "Communication accommodation theory." He explains that while communicating, people try to minimize social differences with the person they are interacting with them. The theory focuses on how humans' tendency of adjusting their behavior while interacting with others. People adjust their communication activities to create a positive image or get approval from the interactant. There are two types of accommodation processes according to this theory: convergence and divergence. In the convergence process, people try to adapt to interactant's communication characteristics to minimize social differences. Whereas in divergence, the process followed is contradictory to the method of adaptation and in the said context, the focus is on social differences between interactants (in Cultural Communication, 2014).

3.3.2 INTERCULTURAL ADAPTATION MODEL

The intercultural adaptation model elaborates on how people adjust in their communication during intercultural communication. The model illustrates how people's previous experiences of intercultural interaction hinder their attempt to adopt intercultural communication (Cai et al., 1997).

3.3.3 RESEARCH QUESTIONS

RQ 1. Identify two popular Indian travel bloggers active on Facebook during the 2020–21 pandemic period.

RQ 2. How did they promote travel destinations during the pandemic lockdown of 2020–21 on their Facebook page?

RQ 3. Identify five popular Indian tourism campaigns on Facebook during the pandemic period of 2020–2021.

RQ 4. How did they promote inter-cultural communication through this campaign?

Objectives:

1. To analyze Indian travel bloggers' pages on Facebook and study their contribution to tourism campaigns during the pandemic period of 2020–21.
2. To study and evaluate how hashtag tourism campaigns on social media promote intercultural communication during the pandemic period of 2020–21.

3.4 RESEARCH METHODOLOGY

To carry out the study, a qualitative analysis was done. Analysis of video posts by two popular Indian travel bloggers on Facebook during the pandemic period of 2020–2021 was carried out. Further, an analysis through the case study of four hashtag travel campaigns by Indian brands on social networking pages to promote intercultural communication during the pandemic period from April 2020 to May 2021 was conducted.

A passive page analysis of the Facebook page of selected Indian travel bloggers (Curly Tales and Tripoto) was done. The period chosen was the pandemic lockdown period in India between May 2020 and August 2020. Content analysis and thematic analysis were conducted on selected targeted posts to get the themes required to understand the objectives (Franz et al., 2019).

3.5 RESULTS AND DISCUSSION

3.5.1 FINDINGS

TABLE 3.1 Facebook Page Content Analysis of Posts of Travel Bloggers (May–August 2020).

Facebook page name	Type of content (video/photo/post)	Message (content-wise)	Hashtag used	The caption of the video/photo (if available)	Period of posting	Total audience engagement in the post (likes +shares + comments)
Curly Tales	video	To give a virtual tour experience of Andamans to viewers. The video shoot before the pandemic was used.	No	I Love My India	7 July 2020	23k+3.4k+ 1.1k
Curly Tales	video	To give a virtual tour experience of Srinagar to viewers. The video was shoot before the pandemic was used.	#ILoveMyIndia	I Love My India	25 July 2020	9.8k+ 1.3k +341
Curly Tales	video	To give a virtual tour experience of Meghalaya to viewers. The video shoot before the pandemic was used.	#spreadpositivity #ArmchairTravel #StayAtHome #StaySafe	I Love My India	2 June 2020	5k+690 +218
Curly Tales	video	To give a virtual tour experience of Havelock Island to viewers. The video was shot before the pandemic was used.	No	I Love My India	14 July 2020	112+13 +9
Curly Tales	video	To give a virtual tour experience of Shillong and surrounding villages to viewers. The video shoot before the pandemic was used.	#spreadpositivity #ArmchairTravel	I Love My India	16 June 2020	19k+5.5k +841
Curly Tales	Post and photo	List of domestic exotic destinations to visit for the next 12 months after lockdown.	#vocalforlocal	12 Destinations in 12 months, here is how you can explore India through the year	27-07-2020	529+110+77

TABLE 3.1 *(Continued)*

Facebook page name	Type of content (video/ photo/post)	Message (content-wise)	Hashtag used	The caption of the video/photo (if available)	Period of posting	Total audience engagement in the post (likes +shares + comments)
Curly Tales	Post and Photo	List of domestic exotic destinations which can be visited after lockdown which are comparable to international travel destinations.	#vocalforlocal	I Love My India: Go #vocalforlocal With These Top 15 International Travel Experiences in India	29-05-2020	4.9k+1k+ 384
Curly Tales	Post and photo	Best international destinations in India which can be enjoyed after lockdown. Emerald lake of Uttarakhand is comparable to the Lake District of the U.K.	#vocalforlocal	no	26-06-2020	1.5k+88+52
Curly Tales	Post and photo	A domestic exotic location worth visiting after lockdown comparing the scenic beauty of Gurudongmar Lake of Sikkim to Iceland lakes.	#vocalforlocal	Here's why Sikkim's Gurudongmar Lake is Better Than Island Lakes?	29-05-2020	746+79+67
Curly Tales	Post and photo	A domestic exotic location worth visiting after lockdown; comparing the scenic beauty of Chitrakoot Falls of Jagdalpur to Niagara Falls of the USA.	#vocalforlocal	no	18-06-2020	2.1k+191+74
Curly Tales	Post and photo	A domestic exotic location worth visiting after lockdown; comparing the scenic beauty of Kaas Valley Plateau of Satara with Lavender Field in Provence of France.	#vocalforlocal	no	16-06-2020	1.2k+68+93

TABLE 3.1 *(Continued)*

Facebook page name	Type of content (video/photo/post)	Message (content-wise)	Hashtag used	The caption of the video/photo (if available)	Period of posting	Total audience engagement in the post (likes +shares + comments)
Curly Tales	Post and photo	A domestic historic location worth visiting after lockdown; comparing the ruins of Rome with Hampi ruins in Karnataka	#vocalforlocal #ArmchairTravel #SpreadPositivity	no	30-05-2020	161+8+7
Curly Tales	Post and photo	A domestic exotic location worth visiting after lockdown; comparing the scenic beauty of Rann of Kutch in Gujarat to Uyuni's Salt Flats in Bolivia.	#vocalforlocal	no	15-06-2020	396+24+8
Curly Tales	Post and photo	A domestic exotic location worth visiting after lockdown; comparing Delhi's India Gate to Paris's Arc De Triomphe.	#vocalforlocal	Forget Paris's Arc De Triomphe! Have You Seen Delhi's India Gate?	10-07-2020	39+1+1
Curly Tales	Post and photo	A domestic exotic location worth visiting after lockdown; comparing the Floating market of Bangkok, Thailand with the Floating market in Srinagar.	#vocalforlocal	I Love My India: Floating Market of Srinagar Is Way Better Than Bangkok's!	1-06-2020	253+15+13
Curly Tales	Post and photo	A domestic exotic location worth visiting after lockdown; Meghalaya touring.	#vocalforlocal #ArmchairTravel #SpreadPositivity #MyBucketList	no	26-05-2020	109+4+20
Curly Tales	Post and photo	Domestic exotic location worth visiting after lockdown; World's biggest Tulip Garden which can be compared with Amsterdam's Tulip Garden opens up in Uttarakhand	#vocalforlocal	World's Biggest Tulip Garden Opens Up in Uttarakhand	21-06-2020	105+8+1

TABLE 3.1 *(Continued)*

Facebook page name	Type of content (video/photo/post)	Message (content-wise)	Hashtag used	The caption of the video/photo (if available)	Period of posting	Total audience engagement in the post (likes +shares + comments)
Curly Tales	Post and photo	A domestic exotic location worth visiting after lockdown; Mumbai's Marine Drive.	#vocalforlocal #ArmchairTravel #SpreadPositivity	no	19-05-2020	786+53+170
Curly Tales	Post and photo	Domestic exotic location worth visiting after lockdown; Radhanagar beach, Andaman is compared with beaches in the Maldives.	no	I Love My India: Here's why Andaman's Radhanagar Beach is Better Than Maldives Beaches.	13-06-2020	112+13+9
Tripoto	Post and photo	A domestic exotic location worth visiting after lockdown and compared with Bali beaches.	#vocalforlocal	Can't Go To Bali? Experience This Stunning Indian Coast On A Roadtrip Instead	8-7-2020	231+61+14
Tripoto	Post and photo	A domestic exotic location worth visiting after lockdown; Pauri Garhwal, Uttarakhand	#vocalforlocal	5 Places in Uttarakhand Perfect for post-pandemic trip #vocalforlocal	14-07-2020	786+149+112
Tripoto	Post and photo	A domestic exotic location worth visiting after lockdown: Scenic beauty of Sikkim is compared with Switzerland	#Vocalforlocal	10 Reasons to go to Sikkim before Switzerland!	1-6-2020	304+53+65

TABLE 3.2 Content Analysis of Comments on the Facebook Posts of Travel Bloggers.

Message	Content format (video/image)	Hashtag used	Initial phase coding	Final phase coding	Category
Virtual tour of footage shoot on exotic domestic location before pandemic	video	#ILoveMyIndia #Spread Positivity #Armchair Travel #StayAtHome #StaySafe	Virtual tour with video footage shoot at the domestic location before the pandemic	Virtual tour with video footage of the domestic location	Virtual tour of the domestic travel destination video
Giving a list of domestic destinations to travel after lockdown and comparing its exotic beauty with foreign popular destinations.	Image and post	#vocalforlocal	A listicle of domestic exotic locations to visit after the pandemic which are no less than popular international destinations.	A listicle of domestic destinations which can be compared with a foreign counterpart	Comparison of the list of domestic locations with foreign destinations.
Comparing a particular domestic destination with a famous foreign destination.	Image and post	#vocalforlocal #Spread Positivity #Armchair Travel #mybucketlist	Re-discovering exotic domestic destinations and comparing them to the foreign popular foreign destinations.	Comparison of exotic or historic domestic location	Re-discovery of domestic destination which is comparable with the foreign popular destination.

Intercultural Communication in Tourism During COVID-19 Pandemic

TABLE 3.3 Thematic Analysis of Comments on Facebook Posts.

Category	Initial theme	Final theme
Virtual tour of the domestic travel destination video	Virtual tour of domestic travel	Promote experience domestic inter-cultural diversity
Comparison of the list of domestic locations with foreign destinations.	Listicle with a comparison between domestic and international destination	Inform, compare and encourage domestic destination travel.
Re-discovery of domestic destination which is comparable with the foreign popular destination.	Rediscovering and comparing domestic destinations with a foreign location	Inform, compare and encourage domestic destination travel.

3.5.2 ANALYSIS

Through content analysis of selected Facebook posts of travel bloggers' (Curly Tales and Tripoto) official pages from May 2020 to August 2020, it was found that some categories of messages were posted to promote the domestic tourism industry once the pandemic-imposed movement restriction on foreign locations came in effect. Bloggers tried to create a desire for visiting local destinations which are at par with any foreign location in their scenic beauty. This awareness cum promotional message through video and images and content posted on their official pages during the lockdown period has contributed to the attempt of the Ministry of Tourism, Government of India's campaign project #vocalforlocal to revive the domestic tourism industry in the post-pandemic lockdown period. Engagement levels in terms of likes, shares, and comments reveal the reach of digital tourism promotion through social media like Facebook. Some location-related posts received more audience engagement in terms of likes, shares, and comments. Exotic locations have higher audience engagement as compared to historical location promotional posts. The pandemic forced the country into lockdown, which directly affected the travel and tourism sector very badly; so, Government of India's initiatives to revive the tourism sector by creating awareness and desire for domestic travel destinations for domestic travelers as the international travel ban was prevalent due to pandemic situation. The content analysis of the posts also showed how some hashtags did repeat posting compared to others. This hashtag created maximum reach for audiences and is promoted as a campaign to revive the tourism sector. #vocalforlocal has maximum usage in a post by Tripoto as well as Curly Tales. Curly Tales has its brand for the promotion of domestic travel under the program name "I Love My India," which shows virtual tours to various locations in India. Most used themes in the posts by both the bloggers are "inform, compare, and domestic destination travel." In almost every post the domestic locations are re-discovered and compared with similar international popular destinations.

3.6   CASE STUDY ANALYSIS

3.6.1   CASE STUDY OF #TRAVELFORINDIA CAMPAIGN

Indian Hotel Company launched the #TravelForIndia campaign on 27th September 2020 on the occasion of World Tourism Day. It is one of the largest tourism companies in South–East Asia. It is an initiative to revive the tourism industry of the country in line with the campaign started by the Govt. Of India "vocal for local" to revive the country's economy (Lifestyledesk, 2020). The first half of the year 2020 was spent in lockdown. After August, the country started reviving back till the first wave hit hard and then again opened to business after January 2021 till the time a second wave hit back in April 2021. Our period of study will be from August 2020 to the beginning of the second-wave April 2021 when the second phase of lockdown started in India.

The central message of the campaign encourages travelers to travel within India, discover various places and boost internal tourism. It projected the beautiful luxurious hotels of the IHCL and how they adhere to the COVID protocol for the safety of their guests. Encourage intercultural communication and experience as India is a culturally diverse nation, so traveling encourages intercultural exploration concerning communication. This in turn enhances local artisan and local tourism-based economy (Lifestyledesk, 2020).

3.6.2   CASE STUDY OF #DEKHOAPNADESH

It is yet another signature campaign of IHCL and promoted in all respect by the brand Taj group of hotels and resorts to promote the internal tourism industry within the country. For one and a half years, there was the pandemic-induced homebound situation, which left a devastating impact on the travel and tourism industry. The two COVID waves further worsened the situation so IHCL decided to further boost the campaign started in 2020 #TravelForIndia and launched its next campaign #DekhoApnaDesh in solidarity with the Ministry of Tourism of the Indian Government attempt to revive the tourism industry. The campaign was conceptualized by the famous advertising agency Rediffusion (BestMediaInfoBureau, 2021).

The central message of this campaign was to call all Indian travelers to explore the cultural diversity of India. The travelers are urged to explore and experience the magnificence of India. IHCL offers through its various brands of hotels like Vivanta, Ginger, and Taj, an experiential stay with windows to cultural and geographical diversity with specially curated itineraries. All these are done to promote local instead of global tourism to boost the tourism industry of India (BestMediaInfoBureau, 2021). Hence, this work is the promotion of intercultural communication with this experiential stay facility in remote, picturesque, and pristine locations of India.

3.6.3   CASE STUDY OF #MYINDIA

MakeMyTrip launched the campaign #MyIndia in early 2021 and gained popularity by showing the hidden beauty of India's geographical diversity by unraveling pristine locales not much visited in past years by homebound tourists. All this was an attempt to lure Indian travelers to domestic travel destinations and revive the domestic tourism industry after the prevailing pandemic situation for more than 1 year. International boundaries were closed to tourists, so the campaign aimed at creating demand among domestic tourists. For this campaign, the travel portal MakeMyTrip collaborated with 22 social media influencers. Each of these influencers took the audiences on a virtual tour of less-traveled destinations explored by each one of them via video, photo, and reels. Each of the posts was meant to generate desire among the viewers for the destination to travel to. Through this campaign, intercultural communication is also encouraged as domestic diversity of destination urges to do so (Newby, 2021).

3.6.4   CASE STUDY OF #REASONTOTRAVEL

Yatra has launched the campaign #ReasonToTravel campaign in August 2020. It is a digital campaign of Yatra.com where the central message is to remind the travelers to start the move as the lockdown has subsided and slowly country is getting back to normalcy. Due to the pandemic lockdown in 2020, travel was almost at a standstill, air travel was shut down, and no tourism-related activity could take place. But after August 2020, the air travel was opening, and life was reviving back to normalcy. The campaign #ReasonTo Travel spread the message about opening up air travel and encourages the customer to start travel for various reasons which include emergency work-related as well as meeting up with family, and relatives to celebrate any occasion (Team, 2020).

3.7 DISCUSSION

Through this extensive study, involving analysis of tourism campaigns by content analysis, thematic analysis, and case study method of several hashtag campaigns launched during this period on various social media platforms by different brands of the tourism sector, it was found that the promotion of domestic tourism is done to the audience via social media during a pandemic lockdown of 2020 and after the lockdown period, too. The promotion focused on the idea of exotic domestic locations, being of an equal standard when compared with international popular destinations of such sorts. The promotion on social networking page by travel bloggers has given a list of such domestic travel destinations with an international feel as far as scenic beauty is concerned. Audience engagement is also studied through each travel blogger's post on their Facebook page.

Toubes et al. (2021) in their research article "Changes in consumption patterns and tourist promotion after the Covid-19 Pandemic" analyzed changes in marketing and promotion in the tourism sector of Spain during and after the pandemic. The research was qualitative where researchers took 65 interviews with experts in the field of tourism and consumer's behavior and marketing. The result showed that online sources of information gained importance as compared to word of mouth from friends and relatives. The digitalization of the tourism sector is likely to displace physical travel agencies. Moreover, AI and VR further added to improvement in online service in the medium-term (Toubes et al., 2021).

Ranasinghe et al. (2020) in their research article "Tourism after corona: Impacts of COVID-19 pandemic and way forward for tourism, hotel and mice industry in Sri Lanka" talked about the economy getting a setback due to the pandemic, which forced the tourism sector in Sri

Lanka to shut down amid growing cases of corona infection in 2020. The researchers tried to find different ways to stabilize this economic instability due to the pandemics. Sri Lanka is an island country, which is majorly dependent on the tourism industry and due to the pandemic, it was severely affected. It requires a way out to stabilize the situation to revive back the growth stage that was there before the pandemic hit it (Ranasinghe et al., 2020).

3.8 CONCLUSION

International tourism often benefits from economic, political, and cultural stability and globalization. Till the pandemic hit the world, intercultural communication was an important part of the tourism industry, both national and international. India is a country of diverse cultures, so traveling to domestic locations will give ample opportunity to explore intercultural diversity and engage in meaningful communication with the locals. This in a way revives local economic growth, which became stagnant due to the pandemic lockdown. The result showed how social media played a very important role in reaching out to an audience who are keen to take travel to domestic locations after unlocking the phase-permitted tourism. The public engagement on each post with mentioned hashtags and themes by bloggers on Facebook has proved public enthusiasm about it. The uniqueness of each domestic location promoted has proved the promotion of intercultural interaction with locals as added experiential advantage in each post.

At the same time, the uniqueness of these not much traveled exotic domestic locations is not only its ample opportunity to experience an intercultural diversity but also the comparing capacity of experience of any foreign destination of such type.

This research is not without limitations. Firstly, it is a recent issue and the pandemic is still not over, so data reliability is challenging as no government source is providing any data. Facebook has several security restrictions, which allow only a few timeline data to be retrieved for a given period. Further research can be done in sentiment analysis of public comments to each travel blogger's post with a similar theme. The quantitative research method can be applied to get more empirical results.

KEYWORDS

- intercultural communication
- tourism campaign
- COVID-19 pandemic
- case study
- Facebook

REFERENCES

Albu, C. E.; Intercultural Communication in Tourism. *Cross Cult. Manag. J.* **2015**, *17* (01), 7–14.

Best Media Info Bureau. *IHCL Invites Travelers to Explore the Undiscovered Beauty of Our Country* [Online], Sep 1, 2021. https://bestmediainfo.com/2021/09/ihcl-invites-travellers-to-explore-the-undiscovered-beauty-of-our-country/

Cai, D. A.; Rodriguez, J. I. *Adjusting to Cultural Differences: The Intercultural Adaptation Model*; 1997.

FICCI Report. *Travel and Tourism - Survive, Revive and Thrive in the Times of COVID-19* [Online], June 2020. https://ficci.in/spdocument/23252/Travel-june-FGT-n.pdf

Franz, D.; Marsh, H. E.; Chen, J. I.; Teo, A. R. Using Facebook for Qualitative Research: A Brief Primer. *J. Med. Internet Res.* **2019**, *21* (8), e13544.

Gulati, S. Social and Sustainable: Exploring Social Media Use for Promoting Sustainable Behavior and Demand Amongst Indian Tourists. *Int. Hosp. Rev.* **2021**. DOI: https://doi.org/10.1108/IHR-12-2020-0072

Communication Theory *Communication Accommodation Theory* [Online], n.d. https://www.communicationtheory.org/communication-accommodation-theory/ (accessed Nov 20, 2021).

Lifestyle Desk. *#travelforindia: IHCL launches campaign to Inspire Safe Travel in India* [Online] Sep 28, 2020. https://indianexpress.com/article/lifestyle/destination-of-the-week/ihcl-launches-campaign-to-inspire-safe-travel-in-india-6619397/

Nair, B. B.; Gastrodiplomacy in Tourism: 'Capturing Hearts and Minds through Stomachs'. *Int. J. Hosp. Tour. Syst.* **2021**, *14* (1), 30–40.

Newby, N. *MakeMyTrip's New Campaign Spotlights India's Hidden Gems* [Online], April 8, 2021. Traveldine. https://www.traveldine.com/makemytrips-new-campaign-spotlights-indias-hidden-gems/

Ranasinghe, R.; Damunupola, A.; Wijesundara, S.; Karunarathna, C.; Nawarathna, D.; Gamage, S.; Idroos, A. A. Tourism After Corona: Impacts of COVID 19 Pandemic and Way Forward for Tourism, Hotel and Mice Industry in Sri Lanka. *Hotel and Mice Industry in Sri Lanka (April 22, 2020)*. 2020.

Team, A. *Yatra.com Launches its Digital Campaign #ReasonToTravel*. Adgully.com – Latest News on Advertising, Marketing, Media, Digital & more [Online], Aug 20, 2020. https://www.adgully.com/yatra-com-launches-its-digital-campaign-reasontotravel-95759.html

Thomas, S.; Kureshi, S.; Yagnik, A. Examining the Effect of Message Characteristics, Popularity, Engagement, and Message Appeals: Evidence from Facebook Corporate Pages of Tourism Organisations. *Int. J. Bus. Emerg. Mark.* **2021,** *13* (1), 30–51.

Toubes, D. R.; Araújo Vila, N.; Fraiz Brea, J. A. Changes in Consumption Patterns and Tourist Promotion after the COVID-19 Pandemic. *J. Theor. Appl. Electron. Commer. Res.* **2021,** *16* (5), 1332–1352.

FACEBOOK PAGES OF BLOGGERS:

Curly Tales Facebook official page - *https*://www.facebook.com/page/124844004231973/search?q=%23vocalforlocal&filters=eyJycF9jcmVhdGlvbl90aW1lOjAiOiJ7XCJuYW11XCI6XCJjcmVhdGlvbl90aW11XCIsXCJhcmdzXCI6XCJ7XFxcInN0YXJ0X3llYXJjcXF-wiOlxcXCIyMDIwXFxcIixcXFwic3RhcnRfbW9udGhcXFwiOlxcXCIyMDIwLTF-cXFwiLFxcXCJlbmRfeWVhclxcXCI6XFxcIjIwMjBcXFwiLFxcXCJlbmRfbW9udGh-cXFwiOlxcXCIyMDIwLTEyXFxcIixcXFwic3RhcnRfZGF5XFxcIjpcXFwiMjAyMC0x-LTFcXFwiLFxcXCJlbmRfZGF5XFxcIjpcXFwiMjAyMC0xMi0zMVxcXCJ9XCJ9In0%3D

Tripoto Facebook official page - https://www.facebook.com/page/321639637922428/search?q=%23vocalforlocal&filters=eyJycF9jcmVhdGlvbl90aW11OjAiOiJ7XCJuYW11XCI6XCJjcmVhdGlvbl90aW11XCIsXCJhcmdzXCI6XCJ7XFxcInN0YXJ0X3llYXJjcXF-wiOlxcXCIyMDIwXFxcIixcXFwic3RhcnRfbW9udGhcXFwiOlxcXCIyMDIwLTF-cXFwiLFxcXCJlbmRfeWVhclxcXCI6XFxcIjIwMjBcXFwiLFxcXCJlbmRfbW9udGh-cXFwiOlxcXCIyMDIwLTEyXFxcIixcXFwic3RhcnRfZGF5XFxcIjpcXFwiMjAyMC0x-LTFcXFwiLFxcXCJlbmRfZGF5XFxcIjpcXFwiMjAyMC0xMi0zMVxcXCJ9XCJ9In0%3D

CHAPTER 4

Deconstructing the Notion of Sacred and Profane from the Viewpoint of Theme-Based Durga Puja in Kolkata in the Age of COVID-19: A Sociological Study

SOUMYA NARAYAN DATTA

Department of Sociology, Bijoy Krishna Girls' College, Howrah, West Bengal, India; Department of Sociology, Adamas University, Kolkata, West Bengal, India

ABSTRACT

Since a few years, Kolkata's Durga Puja has earned the nickname of "Theme Puja." As the term suggests, "Theme Puja" usually points toward those Durga Puja in which the idols are constructed in a nontraditional, that is, not in the traditional or the so-called "sabeki" manner, and the construction of the pandals and the lightings are done on the basis of certain innovative themes. Since its inception, from north to south, such "theme"-based innovations have changed the entire perception of Durga Puja in Kolkata. During the 4 days of the festival, such art depictions surrounding themes have become dominant. However, in the year 2020, apart from diverse themes, certain puja festivals have also come up with themes based on the COVID-19 issue. Based on secondary data, this paper tries to deconstruct the sacred-profane dichotomy initially drawn by Emile Durkheim keeping an eye on the theme of puja festivals in Kolkata in the time of COVID-19.

Language and Cross-Cultural Communication in Travel and Tourism: Strategic Adaptations.
Soumya Sankar Ghosh, Debanjali Roy, Tanmoy Putatunda, & Nilanjan Ray (Eds.)
© 2025 Apple Academic Press, Inc. Co-published with CRC Press (Taylor & Francis)

4.1 EMILE DURKHEIM ON THE SOCIOLOGY OF RELIGION

The sociological definition of religion in the simplest way states that sociological theories and methods are applied toward religious phenomena. History tells us that there has been a very close connection between sociology and the sociology of religion (Cipriani, 2017). The French sociologist Emile Durkheim in "The Elementary Forms of Religious Life" stated that to propound a definition of religion, the first step is to identify the underlying elements of religious life and to look for the common features to be found in all religions. To identify these elements, he stated that the constituent parts of religion must be examined before the system is described produced by their unity (Morrison, 2006). Thus, an adequate definition of religion must take into account the specific elements of religious life before it can elucidate those features common to the whole. Durkheim's search for a definition of religion creates two important directions. The first is the positivistic definition of religion. Durkheim meant by "positive" as the ability to describe religion in terms which are subject to observation, as opposed to speculative thinking (Morrison, 2006). The second constitutes the investigative direction which examines religion by reducing it to elementary parts searching for what is common to all religions. He, then, went on to outline a definition of religion that comprises of two central parts or elementary forms. At first, he stated that entire religions can be defined in terms of a manner of beliefs and rites and secondly, all religions can be defined in terms of their tendency to separate the world into two regions which are the sacred and profane (Morrison, 2006).

4.2 THE SACRED AND PROFANE DISTINCTION WITHIN THE DOMAIN OF DURKHEIM'S ANALYSIS OF RELIGION

It is not possible to gain a proper understanding of Durkheim's theory of religion without elaborating on the difference between the sacred and the profane. This separation of the world into two separate domains of sacred and profane forms one of the central principles of a social theory of religion and in fact, it is the most distinguishable element of religious life. He believed that in several respects, the sacred and profane form the basis of religious life (Morrison, 2006). At first, he contended that the sacred manifests not only Gods, spirits, and natural things but also embraces beliefs. A belief, practice, or rite can possess a sacred character and the

Deconstructing the Notion of Sacred and Profane from the Viewpoint   71

tendency to be viewed by others as a "consecrated thing" makes it sacred. Secondly, words and expressions and even combinations of words can also be sacred (Morrison, 2006). Such words or combinations of words and expressions can be uttered only by consecrated persons and involve gestures and movements that can be performed by specific people only. In addition, a system of rites, beliefs, and social practices that emerges from sacred things radiates around them (Morrison, 2006).

On the other hand, profane is distinct from the sacred. According to Durkheim, in terms of dignity profane is something that is subordinate to the sacred and thus is observed as completely antithetical to the sacred. In this way, profane is the principle that has the capability to defile the sacred and to this scale, the sacred and profane are linked together. Rules exist in every religion that control the separation between the two and safeguards must always be taken when they come into contact (Morrison, 2006). He outlined six characteristics of the sacred and profane. The first characteristic is that the sacred is always segregated from all other objects and hence constitutes things set apart. The second characteristic is that a system of rites and social practices arises that leads to the ways by which the sacred is to be approached and the manner by which the members of the group are to conduct themselves in the presence of the sacred object (Morrison, 2006). Third, sacred things are things protected by proscriptions which have the force of prohibitions or taboos acting to protect and set apart the sacred. Fourth, sacred things are separated from profane things and thus thought to be superior in dignity. Fifth, the sacred and profane represent a principle of unification that separates the natural from the spiritual world and hence provides society with a model of opposites such as good and evil, clean and dirty, holy and defiled, and so on. Finally, entrance from the profane to the sacred must be escorted by rites that are thought to alter one state into the other through rituals of initiation or rebirth (Morrison, 2006).

4.3   DURGA PUJA FESTIVAL—ITS HISTORY AND EMERGENCE

The ceremonial worship of the mother Goddess Durga is regarded as one of the most significant festivals in India. Although it is regarded as a religious festival for the Hindus, it also marks an event for reunion and rejuvenation. Although the rituals entail fast, feast, and worship of the last 10 days, the last 4 days of the festival, that is, Saptami, Ashtami, Navami, and Dashami are celebrated with much festivity and splendor in India and

72 Language and Cross-Cultural Communication in Travel and Tourism

also abroad. Particularly in the state of West Bengal, the 10-armed goddess Durga riding the lion is worshipped with great passion and devotion (Das, 2018).

In West Bengal, a place called Supur situated near Santiniketan, the earliest mention of worship of the Goddess Durga dates to 550 CE (Gupta, 2020). In Markandeya Purana the story of king Surath has been told who lost his kingdom and was compelled to rove for years in the forest. During his exile, King Surath came across another king who was also exiled in the forest named Samadhi who was a Vaishya (Chowdhury, 2014). Thus, both the kings who lost their kingdoms met sage Medha who narrated the story of the emergence of Goddess Durga with 10 hands slaying the buffalo demon, Mahashishasura, and later on suggested both King Surath and Samadhi to invoke Goddess Durga to regain their lost kingdom. Thus, King Surath and Samadhi performed the Basanti Durga Puja on the advice of Sage Medha during the spring season or *Basanta Kaal* which is celebrated in the eastern parts of India and especially in West Bengal (Chowdhury, 2014).

Later on, every year in the month of Ashwin, Durga Puja is celebrated which is also known as Sharodiya Durga Puja, that is, Durga Puja which is celebrated in *Sarat Kaal* or the autumn season. In Ramayana, this Durga Puja during the month of Ashwin commemorates the invocation of the goddess by Lord Rama before going to war with the demon king of Lanka, Ravana to rescue his wife Sita who was abducted and held captive in a garden called Ashok Vatica that was located in the kingdom of Ravana (Das, 2018). Lord Rama first worshipped the Mahishasura Mardini or the slayer of the buffalo demon Mahishasura by offering 108 blue lotuses and lighting 108 lamps at this time of the year. Hence, this autumnal Puja is also known as akal-bodhan or out-of-season ("akal") worship ("bodhan") due to the fact that this autumnal ritual was distinct from the conventional Durga Puja that is usually celebrated in the spring season (Das, 2018).

4.4 DURGA PUJA FROM BEING AN EVENT OF THE LANDLORDS TO BECOMING A COMMUNITY PUJA IN BENGAL

In Bengal's Nadia district, a well-known zamindar of Taherpur, Kansha Narayan instead of the customary Durga Puja in spring was the first to organize the autumn Durga Puja in Bengal. In 1500 CE, Durga Puja was made as an annual affair by Kangsha Narayan and was called the Saradiya

Durgautsav (Gupta, 2020). There is no mention in history to the reason why Kangsha Narayan converted the time of Puja from spring to autumn but a few historians opined that spring being a season of various diseases, for instance, smallpox was unfavorable for such festivals to happen. Thus, autumn regarded as a harvest season was regarded more appropriate for these festivities (Gupta, 2020). To gain popularity among the masses, Kangsha Narayan changed the period of Durga Puja that grew into prominence. In 1606 CE, Bhabananda Majumdar of Krishnanagar in Nadia district who was the ancestor of Krishnanagar's Raja Krishna Chandra Roy started his Durga Puja. Since its inception, the Durga idol that has been worshipped at Raja Krishna Chandra Roy's palace in Krishnanagar is a 15-foot-tall image famously known as Rajrajeswari (Empress) (Gupta, 2020). Lakshmikanta Majumdar who was the founder of the Sabarna Roy Chowdhury family started the first Durga Puja in Kolkata in 1610 CE. Originally, the Sabarna Roy Chowdhury family had landholding rights in the three villages of Gobindapur, Sutanuti, and Kolkata (Gupta, 2020). However, in 1698, the landholding rights for the three villages were acquired by the British from the Sabarna Roy Chowdhury family and carried on to pay regular rent to the Mughal emperor as they had done till the Battle of Plassey in 1757. Later, the colonial rulers developed the three villages to build up the city of Calcutta or Kolkata (Gupta, 2020).

Raja Krishna Chandra Roy initiated the Grand public Durga Puja festivities. In 1757 CE, after Robert Clive's achieved victory in the Battle of Plassey, this Puja became an event of landowners and noblemen. Lord Clive wanted to celebrate his victory over Nawab Siraj-ud-Daulah of Bengal but there was not a church that was left standing in Kolkata. In this situation, Nabakrishna Dev stepped forward. Nabakrishna Dev having a sharp mind deeply understood the British culture (Gupta, 2020). He was proficient in English, Farsi, and Persian languages. In 1750, Nabakrishna Dev who was an enterprising young man was appointed as a Farsi teacher to teach Warren Hastings Farsi language and finish up by helping the colonial rulers win the Battle of Plassey. Subsequently from a Farsi teacher, Nabakrishna went on to become a Munshi, that is, clerk-cum-interpreter of Roger Drake who was the governor of the Council of the Fort William in Bengal from August 1752 to 1758 (Gupta, 2020). In 1756, when Nawab Siraj-ud-Daulah attacked Kolkata, Nabakrishna took the Britisher's side and helped to export food into the city before the city's fall to the Nawab (Gupta, 2020). Later on, Nabakrishna became close to Robert Clive and played an intermediary role in negotiations between the British East India Company and the Nawab of Bengal. Many historians also mentioned that after the Battle of Plassey Nabakrishna was instrumental in helping Clive to steal the treasury of Siraj-ud-Daulah. As a sign of gratitude, British gave the title of "Raja" to Nabakrishna Dev. He gathered huge sums of wealth and acquired the mansion at Sovabazar where he was offered to organize a Durga Puja to celebrate the British victory at the Battle of Plassey (Gupta, 2020). For the grand event, a thakurdalan, or the hall of worship was built. Slaughter of goats was done for a feast and other entertainment was organized to impress the British. After this each year, all other zamindars also started to organize Durga Puja at their own palaces to establish their status. Thus, Durga Puja became an occasion of feasting and merry making from a religious ceremony (Gupta, 2020).

However, the common man had no place to enter in these Pujas. The British were entertained with sherry, champagne, and dance performances by nautch girls. In the late 1700s, a group of young men from the Guptipara village in Hoogly were rudely refused to enter the Durga Puja going on at a local zamindar's grand house. Annoyed at such refusal, these young men decided to start their own Puja (Gupta, 2020). In 1820, a report published in the magazine, *Friends of India* in Serampore stated about a new species of Puja which has been introduced into Bengal within a span of many years known as Barowari. About a number of years ago at Guptipara near Santipura, an association was formed by a number of Brahmins for the celebration of a puja independently of the Shastras rule. As a committee, they elected 12 men from whom it takes its name and sought subscriptions in all surrounding villages. In this way, the *baro-yaar-i*, that is, 12 friends or *Barowari Puja* started in Bengal (Gupta, 2020). Bindhyabasini, a form of the goddess synonymous with Jagadhatri was the deity worshipped during this time. In the magazine, *Friends of India*, it is further mentioned that the worship of Jagadhatri was celebrated for 7 days with much splendor so as to attract the rich from a distance of more than a 100 miles. The procedure of worship was of course regulated by the established practice of Hindu rituals but beyond this, the whole was formed on a plan unrecognized by the Shastras (Gupta, 2020). They acquired the majority of excellent singers to be found in Bengal, entertained every Brahmin who arrived, and spent the entire week with all intoxication of festivity and enjoyment. On the successful ending of the scheme, they determined to render the puja annual and since then it has been celebrated with undeviating regularity (Gupta, 2020).

Deconstructing the Notion of Sacred and Profane from the Viewpoint 75

It is not known for certain about the exact year of the first Barowari Puja. While some sources say that it was 1761, others state that it was 1790. The Barowari Puja that first started in Guptipara started to spread to the neighboring towns of Chinsurah, Santipur, and finally in Kolkata. The community Durga Puja in Kolkata expanded from Barowari Puja to Sarbajanin Puja (everyone's Durga Puja) which became much more elaborate and also expensive (Gupta, 2020). Sudeshna Banerjee mentions in the book, *Durga Puja, Celebrating the Goddess: Then and Now* published in 2006 that probably in 1910, the first Sarbojanin Durga Puja that was held at Balaram Basu Ghat Road in Kolkata was organized by the Sanatan Dharmotsahini Sabha. Soon, local clubs in the areas of Ramdhan Mitra Lane and Sikdar Bagan were holding Durga Puja festivities too. Prominent Durga Puja committees such as the Simla Byayam Samiti and Bagbazar Sarbojanin began to take shape (Gupta, 2020). Thus, it was understood that Durga Puja no longer remained as the preserve of the rich and powerful. In Guptipara there is an area called Bindhyabasini Tala where a temple of Bindhyabasini stands. It is here that a grand Jagadhatri Puja organized by a local puja committee is still celebrated every year (Gupta, 2020). The Barowari puja that originated in Guptipara is no longer the primary festival of the town. Due to its Vaishnava influence, Ratha Yatra is the prime festival that draws huge no. of crowds and thus, the small village of Guptipara is sadly no longer in the spotlight of Barowari Puja (Gupta, 2020).

4.5 RECONSTRUCTING THE SACRED NOTION OF DURKHEIM THROUGH ANALYSING THE THEME PUJAS IN KOLKATA IN THE COVID-19 ERA

Durga Puja as to every Bengali suggests the arrival of the goddess as their parental daughter coming from the abode of Kailash to her paternal home along with her offspring for 5 days that concur with the last 5 days of Navaratri. However, surprisingly, the goddess is worshipped as Mahishasurmardini, or the image showing her as a demon slayer. Riding on a lion as the vahana or vehicle, the goddess slays the demon Mahishasura, or the buffalo demon by piercing the trident into the heart of the demon ("Kolkata's Puja," 2020).

Durga Puja as the Bengali community's mega carnival has of late earned the nickname of "Theme Puja" with different clubs working on different themes to attract the crowd pullers with such diverse innovations. "Theme puja" as the term implies those Durga Pujas in which the idols are constructed in a nontraditional way and the pandals are constructed based on multiple themes. This gives artists an opportunity to show their innovativeness and skills in developing such themes (Mukherjee, 2017). Every year, such unique themes and art not only aim to be unlike but also bring a meaningful message. "Pandal hopping," the word that is used for mass visits to every pandal or marquees during Durga Puja is not only a matter of fun but also a way to comprehend the kind of ideologies that are dominating the present society. Since the time of the emergence of Theme Pujas, social issues have gradually arising as a central theme for many popular pujas in Kolkata (Mukherjee, 2017).

COVID-19 has worsened the lives of people globally making every people to keep confined within their homes. This deadly virus has hardly affected the economy of the entire world. As a preventive measure, while moving outside homes, it has been instructed to follow physical distancing norms, wear masks, and use hand sanitizers to ward off this disease and stay protected. To stop crowding at workplaces—whether in multinational companies, in academic institutions, and also in government sectors—it has been strictly advised to working professionals to stay indoors and go through the work-from-home process. All over India, like other festivals, restrictions have also been imposed in the Durga Puja celebrations across Kolkata. The Honorable High Court stated that visitors will not be allowed to enter inside Durga Puja pandals and so the Puja mandap will be treated as a "containment zone." The Division Bench comprising of Justice Sanjib Banerjee and Justice Arijit Banerjee stated that only the organizers of several Durga Puja committees ranging from 15 to 25 people can enter pandals. The Bench also said that in the year 2020, "No Entry" notices should be placed near the pandals and an awareness drive to inform people about the High Court's order should be taken (Singh, 2020). The Calcutta High Court also said that small Durga Puja pandals and large Puja pandals comprising 5 and 10 m, respectively, should be declared a no-entry zone. A notable advocate appearing for the petitioner also added that the distance from the place where the boundary of the Durga Puja pandal ends will have to be measured. The court also directed that the names of Durga Puja organizers who are allowed inside the puja pandal should be exhibited outside the pandal and this would not change every day (Singh, 2020).

Using the symbolic image of the COVID-19 virus which has been shown in the media since the pandemic struck, several clay artisans of

Kumhartuli, the place located in North Kolkata where most of the clay Durga idols are made on special requests from Puja organizers replaced Mahishasura with "Coronasura" indicating Corona demon by sculpting the head of the so-called depicted demon. While the Durga Puja festivals organized in the neighborhood areas of Kolkata are known for exploring various themes right from fictional to contemporary by getting inspired by traditional art and craft, different kinds of materials are used to construct the marquees and the interior décor which contributes to the cosmopolitan nature of the annual festival, the concept of "Coronasura" may seem bizarre and even frivolous to many and it is not weird for the pandemic to find an expression as a theme in the time of Durga Puja in Kolkata ("Kolkata's Puja," 2020). However, apart from focusing on pandemic themes, there are many puja venues that depicted some unified themes while some other venues have focused on theme-based pandals which have been made after well-known edifices and also some thematic creations on artistic and inspirational ideas are being executed ("Kolkata's Puja," 2020).

HAZRA PARK DURGOTSAB

Hazra Park Puja in South Kolkata was started by Netaji Subhash Chandra Bose in 1942. The year 2020 was the 78th year of the puja. For a long time, Hazra Park puja has maintained its uniqueness even though it did not compete with the other pujas of the state in style. The theme was "Sahajia." It is a fact that in addition to the external form of the organism or the root, a respiratory form is also born within it. That breath is "easy" and that is also easy in children (Biswas, 2020). Thus, to realize this simple truth, Hazra Park is universal and thus has taken shelter with children. Sahajia is neither a community nor a religion. The aim of Hazra Park puja Entrepreneurs was to build this easy vision through their Puja. It is easy to say simple things but it appears to be not easy at all. When the world is grappling with a deadly disease, at such a time the entrepreneurs of Hazra Park want to see Durga Puja with the naked eye. In the wrath of COVID-19, the happy days of childhood are being lost (Biswas, 2020). Thus, the main goal of this puja was to involve children from different walks of life. Instead, they will be given back the light of laughter and education which will help those children to walk healthy in the next life. It is noteworthy that Goddess Durga would be depicted in various children's paintings. Through their paintings, the world of imagination of these children will emerge (Biswas, 2020). Thinking of the children's sufferings in this pandemic situation where they have stopped going to school, and cannot meet their friends physically, they are having a very difficult situation. Artist Krishanu Pal has said that this Puja will be dedicated to these children (Biswas, 2020).

BABUBAGAN CLUB

The organizers of a number of well-known Barowari Durga Puja committees have done research and planned on how to create a "theme" even with an open mandap. In 2020, beyond the conception of the conventional theme, there is a change in theme puja by the puja organizers of Dhakuria's Babubagan Durga Puja, Santosh Mitra Square or Kashi Bose Lane Puja Committee (Feedcrawler user, 2020). The theme of the 59th year Puja of Dhakuria Babubagan is a little village located in the city. Goddess Durga is worshipped in one of the grand halls of that village. In the village environment, cattle carts will be seen. There are scenes that depict rice threshing or ancillary work at home (Feedcrawler user, 2020). Saroj Bhowmik who is on behalf of the Puja Committee and Sujata Gupta who is in charge of planning on mandap creation said, "There is a 35 feet wide road on all four sides of the mandap. The field of worship, the field is absolutely open. However, the mandap seemed to be a bit different from what the theme means."

SANTOSH MITRA SQUARE

Santosh Mitra Square is regarded as one of the most famous pujas held in central Kolkata. This puja completed 65-year in the year 2020. By using the open space of the park, the members of the Puja committee are bringing the Badrinath temple to the premises near Sealdah (Feedcrawler user, 2020). However, the mandap structure appeared to be different. Sajal Ghosh, one of the organizers of the puja, says "Our mandap is open on both sides. The way of entry and exit has been kept the same. However, compared to other times, we have made this path much wider this time. According to the Puja committee, special arrangements have been made for its construction so that the mandap never gets crowded. On behalf of the Puja committee, Puja was to be made live on various social media on behalf of the Puja committee. In a virtual way, those who are unable to come physically will visit the idols sitting at home" (Feedcrawler user, 2020).

(Continued)

KASHI BOSE LANE

Deconstructing the Notion of Sacred and Profane from the Viewpoint 83

The 63rd-year puja of North Kolkata's Kashi Bose Lane in 2020 is Debighat. This puja is one of those who have broken the tradition of theme. The members of the Puja committee agreed that there must be some crowd while worshipping in the narrow alley. However, artist Parimal Pal has kept the four sides of the mandap as well as the roof open so that the visitors do not have to read in a suffocating environment (Feedcrawler user, 2020). Somen Dutt who is the general secretary of the Puja committee said "Even with proper controls on the way in and out, there is still a risk of infection in all closed mandaps." So, we have planned the mandap in a completely different way. No part of the mandap is enclosed. From that point of view, the visitors are much safer in our mandap (Feedcrawler user, 2020).

4.6 RECONSTRUCTING THE PROFANE NOTION OF DURKHEIM THROUGH ANALYSING THE COVID-19-BASED THEME PUJAS IN KOLKATA IN THE COVID-19 ERA

NAKTALA UDAYAN SANGHA

In 2020, Naktala Udayan Sangha in South Kolkata paid tribute to the migrant workers. In this puja, migrant workers are expressed as clouds that go on wandering from one place to the other without any destination. A model of a truck filled with migrant laborers was displayed beside the pandal. Anjan Das who is the additional secretary of Naktala Udayan Sangha stated that the focus of the club was to depict the pain of an employed person. Due to the pandemic, job losses in India have happened severely and unemployment is always at its peak all the time in the country (Datta, 2020). Mr. Das also added that everyone is suffering deeply due to the loss of jobs and pay cuts that COVID-19 has forced upon us. He also pointed out that due to this deadly pandemic, people's lives have been changed radically and the migrant workers, the factory hands, and farmers are those that have suffered the most. These people who are suffering from poverty are struggling for existence (Datta, 2020).

SANTOSHPUR LAKE PALLY

The theme of Santoshpur Lake Pally Puja in the year 2020 was also Corona. However, the image of the demon does not have to look like corona. In this puja, the current situation of corona will be highlighted because, according to the entrepreneurs, this bad time is the real demon. According to the organizers of the Puja committee, millions of people have lost their jobs during the COVID-19 period and for this, the virus is actually responsible (Feedcrawler user, 2020). Hence, human civilization has no demon which is more powerful than corona. According to Somnath Mukherjee, the theme artist of this Puja, "Our country's economy has been hit hard by COVID-19. Many people associated with Puja feel helpless due to such economic conditions. They are seeking the destruction of the demon in the form of a corona in the hands of the goddess. So, in this year's theme, we will paint a picture of the suffering of the common people in the Corona situation" (Feedcrawler user, 2020).

BEHALA 11 PALLY

Behala 11 Pally Sarbajanin in 2020 showed their theme of how Goddess Durga has come to earth and is giving strength to the common people to fight against coronavirus. In the words of entrepreneur Pratip Ghosh, "We have to win the war with Corona. The fight is tough. But victory will be ours. Goddess Durga is coming to stand by our side to increase the morale to fight. That is our theme this year." He said, "We will highlight in the mandap decoration how the goddess tells us to fight against Corona" (Feedcrawler user, 2020).

BARISHA CLUB

In 2020, the Barisha Club Durga Puja committee in the area of Behala in Kolkata portrayed the mother as a migrant worker replaced by the traditional idol of Goddess Durga expressing tribute to their struggles during the COVID-19 pandemic. Idols of Lord Kartik and Ganesh and also goddesses Saraswati and Lakshmi were also shown as idols of migrant workers (Lifestyle Desk, 2020). The female goddesses were replaced with the daughters of the migrant workers carrying duck and owl respectively as their vehicle or "vaahan." There are no weapons nor there is the idol of the demon God Mahishasura. In the wake of the pandemic, the Barisha Club has chosen "relief" as their theme. The idols of the migrant worker have been created to show walking toward the more traditional and smaller image of the goddess having 10 hands inside a nimbus looking for relief (Lifestyle Desk, 2020). Artist Rintu Das said that the goddess is the woman who brazened the burning sun and hunger and beggary along with her children and she is searching for food, water, and some relief for her offspring (Lifestyle Desk, 2020).

4.7 CONCLUSION

In its 16th session from December 13 2021 to December 18, 2021, the intergovernmental committee members of a specialized agency of the United Nations, that is, United Nations Educational, Scientific, and Cultural Organization (UNESCO) focusing on promoting world peace and security through inward collaboration in sciences, arts, and also education and culture on safeguarding intangible Cultural Heritage of Humanity in a virtual meeting held in Paris appreciated Durga Puja for its initiative to involve the-disadvantaged groups and individuals (Saha, 2021). Hence, the iconic Durga Puja of Kolkata that draws tourists from all over the world and is celebrated with great ardor universally has been contained in the Representative List of Intangible Cultural Heritage of Humanity of UNESCO which appears to be a proud moment for every Bengali staying in India and abroad. As a result, at present it is observed that India has 14 intangible cultural heritage elements including Yoga and Kumbh Mela in UNESCO's representative list of intangible cultural heritage (Saha, 2021). After a few months, the West Bengal Government started to plan a major tourism strategy to hardshell the state and make Durga Puja a mega event. A consultant will be appointed by the tourism department to help bring foreign investments aiming to draw more foreign tourists (Gupta, 2022). The consultant would help to identify new areas of tourism to supplement the existing spots and also identify niche concepts such as cultural, heritage, eco, nature, wildlife, wellness, and adventure tourism as well as promoting weekend destinations. Plans are also on the move toward religious tourism circuits and rural tourist spots and also to make Digha, Santiniketan, and Darjeeling special tourism zones. Cultural tourist spots of Baul fakirs such as Gobardanga and Bannabagram are likely to be promoted (Gupta, 2022). The State Government has already announced to make the Durga Puja carnival on Kolkata's Red Road a big event and keeping an eye on this matter, an interactive Durga Puja app is planning to be launched. With the help of the consultant, the tourism department also plans to host roadshows in Delhi and other Indian cities. There are also plans for promotions that could be held abroad (Gupta, 2022).

Drawing on the ideas of the sacred and profane dichotomy that has been described by Emile Durkheim in his book named "The Elementary Forms of Religious Life," in this article, the present author has highlighted upon the fact that theme pujas based on issues around culture, rural life, and certain other issues have been regarded as sacred whereas issues based on the COVID-19 issue focusing on the condition of a migrant worker, the weakening situation of the country's economy that has created a very sharp effect on the professional lives of the people and the people's fight for their existence against corona has been labeled as profane. This indicates that idols and pandals that depict the divine entity of the goddess while observing are termed as sacred whereas the other images and marquees that do not portray the actual divineness that is being observed are designated as profane. Thus, the aforementioned statements stress the fact that the author has moved apart from the sacred and profane perspective drawn by Durkheim, and focusing on the theme of pujas in Kolkata in the COVID-19 era has created a totally different understanding of these ideologies.

KEYWORDS

- **Durga Puja**
- **idol**
- **pandals**
- **sacred profane**
- **COVID-19**

REFERENCES

Biswas, S. Dhormomoter urdhe sohojiyay debi aradhana, sotorkota niye seje uthche Hazra Park. *Eisamay* [Online], Oct 7, 2020. https://eisamay.indiatimes.com/west-bengal-news/

Deconstructing the Notion of Sacred and Profane from the Viewpoint

kolkata-news/hazra-park-durgotsab-committee-celebrating-78-years-of-heritage-puja-this-year-theme-sahajiya/articleshow/78532573.cms

Chowdhury, S. Why is Basanti Durga Puja Celebrated? *Boldsky* [Online], April 4, 2014. https://www.boldsky.com/yoga-spirituality/festivals/2014/why-is-basanti-durga-puja-celebrated/articlecontent-pf35841-039157.html

Cipriani, R. *Sociology of Religion: An Historical Introduction*; Ferrarotti, L., Trans. Routledge, 2017 (Original work published 2000).

Das, S. The History and Origin of the Durga Puja Festival. *Learn Religions* [Online], Sep 3, 2018. https://www.learnreligions.com/the-history-and-origin-of-durga-puja-1770159

Datta, S. Pandemic Themes Reign at Kolkata Durga Pujas. *Zenger* [Online], Oct 25, 2020. https://www.zenger.news/2020/10/25/pandemic-themes-reign-at-kolkata-durga-pujas/

Feedcrawler user. Coronamoy themey kothao osur virus, kothao manusher kosto. *Eisamay* [Online], Sep 3, 2020. https://eisamay.indiatimes.com/west-bengal-news/kolkata-news/this-year-durga-puja-will-see-corona-themes-and-suffering-of-people-due-to-the-pandemic/articleshow/77894532.cms

Feedcrawler user. Nojor jonoswasthey, themey thekeo tayi themer protha bhangar pujo. *Eisamay* [Online], Oct 17, 2020. https://eisamay.indiatimes.com/west-bengal-news/kolkata-news/this-year-durga-puja-theme-focus-is-on-public-health/articleshow/78704514.cms

Gupta, A. Guptipara: Birthplace of the People's Durga Puja. *Live History India* [Online], Oct 22, 2020. https://www.livehistoryindia.com/story/history-daily/durga-puja-guptipara/

Gupta, R. State Eyes Tourism Push via Consultant. *The Times of India*, Feb 17, 2022. https://timesofindia.indiatimes.com/city/kolkata/state-eyes-tourism-push-via-consultant/articleshow/89627249.cms

Kolkata's Puja Organisers Focus on Pandemic Related Themes This Year. *Outlooktraveller* [Online], Oct 17, 2020. https://www.outlookindia.com/outlooktraveller/travelnews/story/70734/pandemic-related-themes-rule-kolkata-durga-puja-this-year

Lifestyle Desk. A Durga Puja Pandal Showcases Women Migrant Workers in Place of the Goddess. *The Indian Express*, Oct 22, 2020. https://indianexpress.com/article/lifestyle/life-style/durga-puja-pandal-to-showcase-women-migrant-workers-in-place-of-goddess-6757447/

Morrison, K. *Marx, Durkheim, Weber: Formations of Modern Social Thought*, 2nd ed.; Sage Publications, 2006.

Mukherjee, B. Theme, a Topic that Dominates Durga Puja Every Time in Bengal. *United News of India*, Aug 31, 2017. https://www.uniindia.com/news/other/theme-a-topic-that-dominates-durga-puja-every-time-in-bengal/975629.html

Saha, S. Durga Puja gets UNESCO tag: What is Intangible Cultural Heritage and How New Status will Help in Its Protection and Promotion. *Financial Express,* Dec 17, 2021. https://www.financialexpress.com/lifestyle/durga-puja-gets-unesco-tag-what-is-intangible-cultural-heritage-and-how-new-status-will-help-in-its-protection-and-promotion/2382333/

Singh, S. S. Coronavirus| No Visitors Allowed Inside Durga Puja Pandals Across West Bengal: Calcutta High Court. *The Hindu*, Oct 20, 2020. https://www.thehindu.com/news/cities/Kolkata/coronavirus-no-visitors-allowed-inside-durga-puja-pandals-across-west-bengal-calcutta-high-court/article32892114.ece

CHAPTER 5

COVID-19 Pandemic and the End of Overtourism: A Perspective

RAJDEEP DEB and PANKAJ KUMAR

Department of Tourism & Hospitality Management, Mizoram University, Mizoram, India

ABSTRACT

The COVID-19 pandemic has brought the economic activities over the world to a halt to an unprecedented degree. Despite the intent of government to revive the tourism sector, too much cannot be expected too soon. The current chapter is a manifestation of the authors' views on the role of coronavirus in shutting the mainstream tourism activities, due to havoc created by COVID-19, and thereby reducing profoundly the footfalls in tourist destinations. In particular, COVID-19 has led to the creation of a space in which there exists an opportunity for escaping overtourism and also consolidating the transition toward an eco-friendly and more responsible tourism. The chapter relied essentially on qualitative research and was mainly descriptive in nature. Finally, the current chapter is also an attempt to put forth how academicians and policy makers can play a pivotal role in current crisis by reconsidering and restructuring the present curriculum and policies to get future generations ready to adopt more greener and responsible travel and tourism practices.

5.1 INTRODUCTION

The term overtourism happened to be the travel buzzword of 2019, as destinations all over the world wrestled the impact of excessive tourist

---

Language and Cross-Cultural Communication in Travel and Tourism: Strategic Adaptations.
Soumya Sankar Ghosh, Debanjali Roy, Tanmoy Putatunda, & Nilanjan Ray (Eds.)
© 2025 Apple Academic Press, Inc. Co-published with CRC Press (Taylor & Francis)

92   Language and Cross-Cultural Communication in Travel and Tourism

trails. Undoubtedly, the contribution of tourism to job creation is staggering, but at the same time the environmental consequences of tourism is also difficult to ignore. The concept of "overtourism" refers to the uncurbed swelling of visitors resulting in overcrowding and agony of local populace, due to interim and often seasonal tourism heights, causing enduring changes in way of living, amenities, and prosperity. Moreover, overtourism-generated complexities in destinations have provided impetus to the strengthening of policy making and scholarly consideration toward finding out solutions to an issue that is regarded as bewildering and troublesome. At the starting of the 21st century, experts were of the opinion that by 2020, world tourism would achieve 1 billion international arrivals per year. The number seemed absurd at that moment, but the target was achieved much before schedule. Moreover, WTTC showed a record 1.5 billion international arrivals in 2019. The unchecked growth of tourism across the world has brought about an unjust and unsustainable earth as well as has germinated anguish among the local communities. The growth of tourism activities (Butowski, 2019) and tourist arrivals in few particular destinations has led to overcrowding and issues with the carrying capacity (Schneider, 1978) and environmental sustainability (Kowalczyk, 2010).

The COVID-19 pandemic has brought the economic activities over the world to a halt to an unprecedented degree. The scope as well as effects of this lockdown triggered by the corona terror worldwide has genuinely confused tourism academicians, policy makers, and practitioners (Miles and Shipway, 2020). While focusing on exploration of recovery affairs of contemporary crises, there is a huge possibility that tourism industry will exhibit its remarkable resilience to this unanticipated market upset, predominantly because of state involvement in the form of stimulus packages. Despite the intent of government to revive the tourism sector, too much cannot be expected too soon. However, the silver lining in such a crisis is that it offers the much-needed opportunity to ponder over a novel direction and move ahead by hugging a more sustainable route. The current chapter is a manifestation of the authors' views on the role of coronavirus in shutting the mainstream tourism activities, due to havoc created by COVID-19, and thereby reducing profoundly the footfalls in tourist destinations. In particular, COVID-19 has led to the creation of a space in which there exists an opportunity for escaping overtourism and also consolidating the transition toward an eco-friendly and more responsible tourism. Taking the vast scope of the subject, the paper largely focused on

COVID-19 Pandemic and the End of Overtourism 93

creating a perspective about the end of overtourism on account of the onset of COVID-19 pandemic.

5.2 LITERATURE REVIEW

Although the term "overtourism" is relatively a new concept and an emerging subject of scientific research, the literature is filled with profound definitions of the phenomenon (Koens et al., 2018; Zemla, 2019; Mihalic, 2020). However, almost all authors hold that it is quite complex to reach consensus about the conceptualization of overtourism. Koens et al. (2018) assert that the emergence of overtourism can largely be bestowed upon media-induced discourses without having any kind of concrete theoretical framework. According to Capocchi et al. (2019), the new concept lacks clarity as it was introduced without being properly defined, thus creating nuisances for the operationalization of the definition. This makes the term overtourism a multidimensional concept (Koens et al., 2018). Conceptually, overtourism relates to two different, but related, perspectives. The first one concerns (negative) experiences of resident population and visitors, whereas the second relates to thresholds for the carrying capacity of destinations (Nilsson, 2020). Table 5.1 lists out some of the definitions of overtourism that were cited in the literature.

The above delineation puts forth the fact that overtourism occurs in a situation where the number of tourists surpasses the carrying capacity of the destination. The World Travel & Tourism Council in cooperation with McKinsey & Company conducted a survey on the consequences of overtourism and found five challenges associated with it. They are—estranged local residents, worsened tourist experience, strained infrastructure, environmental degradation, and cultural and heritage hazard. Overtourism is not only about how much tourism takes place in a destination, but is perhaps even more particularly about what kind of tourism happens, how it does, and how it is being managed (UNTWO, 2019). Higgins-Desbiolles et al. (2019) see overtourism-induced problem as a writing on the wall about how these damages the social and economic optimism stimulated by the tourism sector (Alonso-Almeida et al., 2019).

While considerable research in this domain has concentrated on the density of the visitors from the standpoint of tourists, "overtourism" is a phenomenon is closely connected to the residents' perspectives (Koens

94 Language and Cross-Cultural Communication in Travel and Tourism

TABLE 5.1 Overtourism Definitions Presented in the Literature.

Author and year	Definition	Factors defining overtourism
Goodwin (2017)	Overtourism is a situation wherein every party develops a notion that there are excessive individuals visiting a particular destination and that the goodness of life in the destination or the goodness of the experience has worsened considerably.	Overcrowding, adverse effects on citizens' quality of life, adverse effects in tourists' experiences
Butler (2018)	Overtourism directs toward a situation in which many visitors overload the services and facilities available and also become a serious consequential for permanent residents of these locations.	Impacts, exceeding capacities
Peeters et al. (2018)	Overtourism brings forth a situation where in the influence of tourism, at certain times and in certain places, surpasses physical, environmental, social, economic, psychological, and/or political carrying capacity.	Impacts, exceeding capacities
UNWTO (2018)	A circumstance in which the effect of tourism on a certain destination or parts thereof, massively impacts the discerned standard of life of residents and/or visitors in an adverse way.	Consequences, adverse influences on residents' standard of life
Higgins-Desbiolles et al. (2019)	Overtourism delineates a particular situation in which a tourism destination overshoots its capacity thresholds—in physical and/or psychological terms.	Surpassing capacities
Milano et al. (2019)	The unstopped flow of visitors often results to overcrowding in locations where residents face the ill effects of temporary and seasonal tourism bursts, which have induced permanent changes to their lifestyles, denied access to amenities, and injured their general well-being.	Overcrowding, adverse effects on citizens' quality of life
Perkumiene et al. (2019)	Overtourism is a state which is characterized by too many visitors, which has negative consequences on the quality of the region.	Overcrowding, adverse effects on citizens' quality of life

Source: Adapted from Zemla (2019)

et al., 2018; Gossling et al., 2020). Studies exhibit that crowding sets in a change in the perception of the liveability, desirability, or economic viability of a destination (Lawson et al., 1998; Bellini et al., 2017). Traditionally, overtourism has got mentions specifically in urban contexts (Koens et al., 2018), but similar occurrence can also be extended to other entities, particularly amenity centers and other attractions (Milano et al.,

2019). According to Peeters et al. (2018), negative outcomes can be linked to economic, social, or environmental concerns (Weber et al., 2017).

The aforesaid discussion brings forth the side-effects of overtourism as well as the concern that academics and practitioners have displayed toward this social and environmental menace. In fact, it has become one of the key problems for the areas that have been at the top of the itineraries of visitors. Also, this is becoming a bottleneck to the long-term prospects of tourism development. Hence, it is of utmost significance to explore ways to control tourist numbers and thereby avoiding overtourism.

5.3 MATERIALS AND METHODS

The chapter relied essentially on qualitative research and was mainly descriptive in nature. The information was mainly drawn from secondary sources, particularly journals, books, government published reports, newspapers, etc.

5.4 RESULTS AND DISCUSSION

The world's tourism is growing rapidly with a greater number of people undertaking national and international travel for the first time. With an increasing demand, several destinations have been found to be ill equipped to deal with the economic, social, and cultural consequences of overtourism. Several cities recorded economic growth fueled by consumption of travelers but at the cost of quality of life for locals. Hence, it was quintessential to be highly serious about how and when one undertakes travel in order to avoid stringent rules and restrictions. Local authorities have been limiting the number of tourists and executing rules to make sure that there are plans to safeguard workers, flora and fauna, and popular destinations. Also, destinations all over the world have been contemplating actions to overturn the impacts of overtourism.

Prior to the onset of COVID-19 pandemic, urban tourism witnessed a significant rise across Europe thus leading to debates over carrying capacity and "overtourism" (Goodwin, 2017). The COVID-19 has relentlessly wreaked havoc across different industries especially tourism sector, but still there exists a silver lining in the midst of this catastrophe. It has indeed provided a scope to critically evaluate the existing tourism policies and plan, and a ray of hope that the way forward will auger a change in the way countries manages overtourism. The change has already caught the attention of authorities as there shows a positive consequence of people staying inside their home on the planet already hurt by global warming and uncontrolled tourism. The destinations that, before COVID-19 pandemic, struggled with too many visitors left reeling post the striking of catastrophe. The flying surge in the outspread of the pandemic brought the whole world to a halt and the tourists were forced to come into curfew or lockdown mode and thereby pressing the brakes on their movement. The enforcement of quarantine brought an immediate halt on all kinds of commercial activities that are considered to be the biggest culprits in deteriorating the environmental parameters, which are directly or indirectly associated with human and societal health. As all kinds of economic, social, industrial, and urbanization activity suddenly closed, nature got an opportunity to recover and revamp itself. It seems nature is all but ready to press the reset key, claiming back the encroached spaces to core itself as the human-related activities have declined. However, between all the pessimism that has been generated by COVID-19 pandemic, there is likely to be a break in the clouds so far as restoration of environmental health is concerned. Moreover, it begun to show symptoms of improvement in terms of air quality, cleaner rivers, declined noise pollution, tranquil, and undisturbed wildlife (Fig. 5.1).

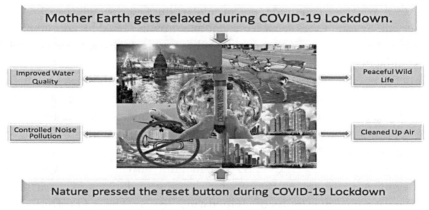

FIGURE 5.1 Healing of earth during COVID-19 lockdown.

Source: Reprinted with permission from Arora et al. (2020) Copyright © 2020 Elsevier B.V. All rights reserved.

COVID-19 Pandemic and the End of Overtourism

The COVID-19 pandemic, according to the published data of UNWTO, has triggered a fall of 22% in international tourist arrivals during the quarter one of 2020. Furthermore, the current predicament could fuel an annual decline of 60 and 80% when measured with the figures of 2019. The tourist arrivals in March 2020 fell considerably by 57% because of the deployment of lockdown in several countries as well as the enforcement of travel limitations, shutdown of airports, and closures of national borders. The current pandemic has compelled the travelers not to venture outside and literally forced them to stay at home. Figure 5.2 displays the percentage change of the international tourist arrivals till September 2021 over the year 2019.

The COVID-19 pandemic has in many ways offered a weapon in the hands of state machinery to combat the menace of overtourism. It has brought a halt to the travel across the globe, which is in a way directly or indirectly disrupting normal life and environmental balance. The restriction on air travel across the globe is welcome news for the environment since aviation alone is responsible for nearly 2% of worldwide greenhouse gas discharge.

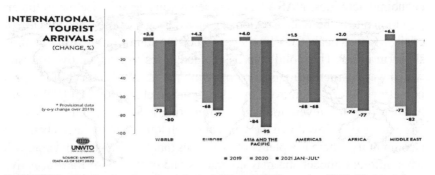

FIGURE 5.2 International tourist arrivals, 2019 and till September 2021 (% change). Source: UNWTO; * Change over 2019 (provisional data).

Going by the prediction of the World Tourism Organization (UNWTO), international tourist arrivals will jump considerably by 2030 (Fig. 5.3).

The illustration given in Figure 5.3 presents the anticipated tourist flows for the next 10 years. This indeed is going to be a concern for the world communities as it may put tremendous pressure on the earth and challenge the sustainability. According to the data published by the French Environment and Energy Management Agency (ADEME), it may be inferred that a kilometer by plane is 45 times more polluting than a kilometer on

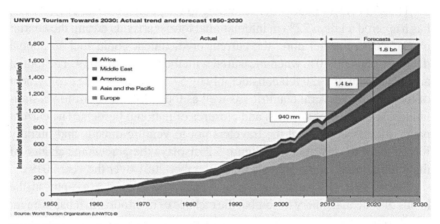

FIGURE 5.3 International tourist arrivals by 2030.
Source: UNWTO Tourism Highlights, 2017 Edition

a high-speed train. Due to the stoppage of air flights along with other economic activities, NASA has recorded a considerable decline in the air pollution rate in the Wuhan region of China during the time when the current crisis was at its peak. COVID-19 has forced the authorities to restrict tourist shops in cities. This solid step has helped cities to manage overtourism, which were earlier struggling to cope with the negative impact of rampant homestay culture. Moreover, by restricting travel and tourism COVID-19 has managed to prevent tourism from injuring local communities. Also, it has managed to slow down the speed at which tourism stimulated change in a community. There is a plenty of evidences that COVID-19 pandemic will have different impact and may emerge as one of the major transformative forces for the tourism sector. From the perspective of tourism, this current pandemic may be viewed as an opportunity to escape and beat overtourism and transforming the existing business model into a long-term sustainable and environment friendly industry. Travelers in all probability will be interested towards limited and more worthy trips. In fact, they will try to avoid the destinations which are likely to remain overcrowded.

5.5 CONCLUSIONS AND IMPLICATIONS

Although the economic and social impact of the COVID-19 pandemic on the world's significant markets is still unclear, the positive side is that it managed to heal the ailing earth from the constant pressure of overtourism.

COVID-19 Pandemic and the End of Overtourism

Prior to the onset of COVID-19 pandemic, major tourist destinations were overburdened with visitors, which had a tremendous long-term adverse impact on the environmental health. The current crisis has presented the world with an opportunity to salvage the earth by developing a smaller, slower, and more sustainable model of world tourism. The incumbent situation has greatly revamped the worsening environmental health and thus made it the extent of adverse impact the travel and tourism have on the planet earth. With world gradually returning to normal, there needs a concerted efforts from several stakeholders to safeguard not only humans but also the earth, to stop the emergence of catastrophes like revenge tourism. So, the prevalent situation may create a scope for green tourism to flourish where offbeat destinations, the wilderness, and home stay in countryside or villages may be preferred to big budget hotels that lead to carbon footprints. Further, the crisis has also created a scope for discussion on the necessity of reassessing and reconstructing future tourism policies. Tourism is at a crossroads and the restorative steps taken today will give a shape to the tourism of tomorrow. Also, governments need to take into consideration the long-term effects of the crisis, while focusing on digitalization, underpinning the low carbon release, and nurturing the structural changes required to construct a robust, more sustainable, and buoyant tourism economy. The pandemic has laid bare the skeleton of the interplay between tourism, delicate environment, and community like never before. It strongly offers an opportunity to speed up sustainable consumption and production patterns and rejuvenate better tourism.

Further, over the past many years, the term "cheap or affordable" has instigated tourism growth and thereby creating an ecosystem for overtourism across the globe, as the number of tourists outweighs the carrying capacity of tourist destinations. With the commoditization of destinations in order to attract a greater number of tourists to support local economies, it has exposed destinations to disruptions in tourist flows. Moreover, the mass tourism has made several destinations vulnerable to considerable environmental affects in terms of air, water, soil, and noise pollution and thereby intimidating the ailing ecosystems.

Lastly, the current chapter is also an attempt to put forth how academicians and policy makers can play a pivotal role in current crisis by reconsidering and restructuring the present curriculum and policies to get future generations ready to adopt more greener and responsible travel and tourism practices.

KEYWORDS

- **COVID-19 pandemic**
- **crisis**
- **ending**
- **overtourism**

REFERENCES

Alonso-Almeida, M. D. M.; Borrajo-Millán, F.; Yi, L. Are Social Media Data Pushing Overtourism? The Case of Barcelona and Chinese Tourists. *Sustainability* **2019**, *11* (12), 3356. DOI: 10.3390/su11123356

Arora, S.; Bhaukhandi, K. D.; Mishra, P. K. Coronavirus Lockdown Helped the Environment to Bounce Back. *Sci. Total Environ.* **2020**, *742*, 140573. DOI: 10.1016/j. scitotenv.2020.140573

Bellini, N.; Go, F. M.; Pasquinelli, C. Urban Tourism and City Development: Notes for an Integrated Policy Agenda. In: *Tourism in the City: Towards an integrative agenda on urban tourism*; Bellini, N., Pasquinelli, C., Eds.; Springer International Publishing: Switzerland, 2017.

Butler, R. Challenges and Opportunities. *World. Hosp. Tour. Them.* **2018**, *10* (6), 635–641. http://www.emeraldinsight.com/loi/whatt

Butowski, L. Tourist Sustainability of Destination as a Measure of Its Development. *Curr. Issues Tour.* **2019**, *22*, 1043–1061. DOI:10.1080/13683500.2017.1351926

Capocchi, A.; Vallone, C.; Pierotti, M.; Amaduzzi, A. Overtourism: A Literature Review to Assess Implications and Future Perspectives. *Sustainability* **2019**, *11* (12), 3303. DOI: 10.3390/su11123303

Goodwin, H. *The Challenge of Overtourism*; Working Paper 4; Responsible Tourism Partnership: Cape Town, 2017.

Gössling, S.; McCabe, S.; Chen, N. C. A Socio-Psychological Conceptualisation of Overtourism. *Ann. Tour. Res.* **2020**, *84*, 102976. DOI: 10.1016/j.annals.2020.102976

Higgins-Desbiolles, F.; Carnicelli, S.; Krolikowski, C.; Wijesinghe, G.; Boluk, K. Degrowing Tourism: Rethinking Tourism. *J. Sustain. Tour.* **2019**, *27* (2), 1926–1944. DOI: 10.1080/09669582.2019.1601732

Koens, K.; Postma, A.; Papp. B. Is Overtourism Overused? Understanding the Impact of Tourism in a City Context. *Sustainability* **2018**, *10* (12), 4384. DOI: 10.3390/su10124384

Lawson, R. W.; Williams, J.; Young, T. A. C. J.; Cossens, J. A Comparison of Residents' Attitudes Towards Tourism in 10 New Zealand Destinations. *Tour. Manage.* **1998**, *19*(3), 247–256. https://www.academia.edu/34797022/A_comparison_of_residents_attitudes_towards_tourism_in_10_New_Zealand_destinations.

COVID-19 Pandemic and the End of Overtourism

Mihalic, T. Conceptualising Overtourism: A Sustainability Approach. *Ann. Tour. Res.* **2020**, *84*, 1–12. DOI: 110.1016/j.annals.2020.103025

Milano, C.; Novelli, M.; Cheer, J. M. Overtourism and Tourismphobia: A Journey Through Four Decades of Tourism Development, Planning and Local Concerns. *Tour. Planning Dev.* **2019**, *16* (4), 353–357. DOI: 10.1080/21568316.2019.1599604

Miles, L.; Shipway, R. Exploring the COVID-19 Pandemic as a Catalyst for Stimulating Future Research Agendas for Managing Crises and Disasters at International Sport Events. *Event Manage.* **2020**, *24*, 537–552. DOI: 10.3727/152599519X15506259856688

Koens, K.; Postma, A.; Papp, B. Is Overtourism Overused? Understanding the Impact of Tourism in a City Context. *Sustainability* **2018**, *10* (12), 4384. DOI: 10.3390/su10124384

Kowalczyk, A. *Turystyka Zrównowazóna*; Wydawnictwo Naukowe PWN: Warszawa, 2010.

Nilsson, J. H. Conceptualizing and Contextualizing Overtourism: The Dynamics of Accelerating Urban Tourism. *Int. J. Tour. Cities* **2020**, *6* (4), 657–671. DOI: 10.1108/IJTC-08-2019-0117

Peeters, P. M.; Gössling, S.; Klijs, J.; Milano, C.; Novelli, M.; Dijkmans, C. H. S.; Mitas, O. *Research for TRAN Committee-Overtourism: Impact and Possible Policy Responses*; European Parliament, Directorate General for Internal Policies, Policy Department B: Structural and Cohesion Policies, Transport and Tourism: Brussels, Belgium, 2018.

Perkumiene, D.; Pranskuniene, R. Overtourism: Between the Right to Travel and Residents' Rights. *Sustainability* **2019**, *11* (7), 2138. DOI: 10.3390/su11072138

Schneider, D. *The Carrying Capacity Concept as a Planning Tool*; American Planning Association: Chicago, 1978.

UNWTO Tourism Highlights. 2017. https://www.e-unwto.org/doi/pdf/10.18111/9789284419029 (accessed on Jan 2, 2022).

UNWTO. *Overtourism? Understanding and Managing Urban Tourism Growth Beyond Perceptions: Executive Summary*; UNWTO: Madrid, 2018.

UNWTO *"Overtourism"? Understanding and Managing Urban Tourism Growth Beyond Perceptions: Executive Summary*; World Tourism Organization: Madrid, 2019.

Weber, F.; Stettler, J.; Priskin, J.; Rosenberg-Taufer, B.; Ponnapureddy, S.; Fux, S. et al. *Lucerne University of Applied Sciences and Arts*; Tourism Destinations Under Pressure. Challenges and Innovative Solutions; Lucerne, 2007.

Zemla, M. Reasons and Consequences of Overtourism in Contemporary Cities—Knowledge Gaps and Future Research. *Sustainability* **2020**, *12* (5), 1729. DOI: 10.3390/su12051729.

CHAPTER 6

Transfiguring the Troubled Past Through Narratives

DUYGU ONAY ÇÖKER

Department of Visual Communication Design, Faculty of Architecture and Design, TED University, Turkey

ABSTRACT

Is it possible to have an ethical encounter for the two sides of a battle sharing a troubled past? This chapter argues yes; even if the battle is a founding event of the national discourse for both sides as in Gallipoli. As an invitation of welcoming and taking responsibility in sympathy for the story of the other, this chapter focuses on a dream of a woman who travels with historical others. Applying the perspective of French Philosopher Paul Ricoeur, this chapter argues that an ethical engagement is possible building a hermeneutical bridge. In order to interpret otherness, this chapter uses the method of recounting the historical narratives by exchanging stories that allows the possibility of seeing each other through the eyes of the other.

6.1 INTRODUCTION

The Gallipoli Campaign took place in Gallipoli, Turkey, during World War I in 1915–1916, and was fought between the Allied Powers and the Ottoman Empire. This study focuses on the historical narratives of both sides of the battle, and discusses the possibility of an ethical encounter years later from the perspective of French Philosopher Paul Ricoeur.

Language and Cross-Cultural Communication in Travel and Tourism: Strategic Adaptations.
Soumya Sankar Ghosh, Debanjali Roy, Tanmoy Putatunda, & Nilanjan Ray (Eds.)
© 2025 Apple Academic Press, Inc. Co-published with CRC Press (Taylor & Francis)

The main objective of this chapter is to discuss the questions "Does Paul Ricoeur's ethical philosophy provide an ethical encounter for understanding the historical narratives of others?" and "Is building a hermeneutical bridge to understand the other possible?" For this purpose, this chapter applies Ricoeur's ethical philosophy to a historical story, through the narratives of the historical others. Therefore, it uses the ethical philosophy of Ricoeur and applies his theory of linguistic hospitality with its narrative plurality, flexibility, and transfiguring of the past.

For this purpose, this chapter narrates an original story of an unnamed woman living in the present time in Gallipoli, Turkey, who reads books about the Gallipoli Campaign. She reads narratives written by one who was accepted as the enemy—the other in the Gallipoli Campaign, where her grandfather also fought on the Turkish side. As the story continues, the theory of Ricoeur is intertwined with the story, and the philosopher's teaching is discussed in the context of the story.

The possibilities of revisiting every story, which has been handed down and of carving out a place for several stories, directed towards the same past.

Paul Ricoeur
Reflections of a New Ethos for Europe

6.2   BACKGROUND OF STORY

This chapter, focusing on a dream of a woman, discusses the possibilities of an ethical encounter.[1] Reflecting upon this encounter, this paper uses not only the stories from her great grandfather but she buries herself in the stories, the still-living narratives of the participants, the stories that Australian writers (Scates et al., 2015) collect together a century later about the Gallipoli battle and the reflections of the French philosopher Paul Ricoeur, that acts of interpreting, imagining, narrating, and recounting should

---

[1]An unnamed Turkish woman, living in Gallipoli, Turkey, reads a book on the Gallipoli battle by Australian writers and falls asleep. In her dream, she encounters some of the Australian soldiers, protagonists of the book she was reading. Through her dream, meeting the soldiers one by one, listening to their stories, she transfigures the past, recognizes, understands, and remembers the other without demonizing her/him, and reads the line between the self and the other. Her place has a beautiful view of Arı Burnu historical memorial of the Gallipoli battle and she has a special relationship with the history of Gallipoli, where her grandparents lived the cruel reality of the war.

Transfiguring the Troubled Past Through Narratives 105

be open to the story of the other, which is achieved through linguistic hospitality with its narrative plurality and flexibility, and transfiguring the past (Ricoeur, 2006, pp. 11–50). Through them, this chapter focuses on building a hermeneutical bridge that includes the other's narrative as a way of thinking to understand the other. The purpose of remembering the past, welcoming the stories of the others, is that it provides her a way of bringing history to life. An intriguing way of making her present in a different way when relating to history. This transfiguring "… involves a creative retrieval of the betrayed promises of history. It permits us, for example, to respond to our 'debt to the dead' and endeavor to give them a voice. The goal of tolerant testimonies is, therefore, 'to try to give a future to the past by remembering it in a more attentive way, both ethically and poetically.'" (Kearney, 2007, p. 157). The past is not only what is bygone-that which has taken place and can no longer be changed. On the contrary, Gallipoli battle is the founding event of a national discourse.

6.3 STORY AND DISCUSSION

Her train was the 5.20 am from Gallipoli, the morning still dark and chilly. Waiting for departure time was always stressful for her: this is why she had arrived at the station at the last minute. She tried not to think about leaving her beautiful town, struggling with thoughts of her return. She heard the train whistle, and opened the book she had brought with her. It was her only piece of luggage, no suitcase, no handbag.

It seemed that she would be alone in the compartment during the journey, a great opportunity for reading. Her book, of war stories, was a family heirloom. She thought this book could reduce her anxiety about her journey because its protagonists were real people, like her great –grandfather, who had fought at Gallipoli, and because she knew that some of the stories were about the battle fought there in 1915. She was born and grew up in Galli-poli. Since she had never left her village before, she was both excited and nervous about travelling to uncertainty—even if the place she would be going to visit was familiar. Because her great grandfather and all his generation had fought in the Gallipoli war, they had told many stories about that time. And of course, women from both sides had lived and narrated the heartbreak they had felt. Through the stories of heroism and suffering she felt connected, not just to her own people's history but also to the other side. Travelling to a strangely familiar land to see the hometown of these heroic people with her own eyes, though it would take a few days, might calm her anxiety. She felt the train starting to move, opened her book, and heard somebody coming into her compartment. Somebody, with an unfamiliar military uniform that she did not recognize: a figure of alterity, a stranger/other, an evil presence, an angel? A figure she could already see that she would have to face, confront, make a decision about.

Self begins journeys through others, since self is not transparent to itself. "The shortest route from self to self is through the other" Richard Kearney (2007, p. 1) writes. The other and the other people,[2] the strange and the familiar, pervade this narrative of an unnamed woman brought face-to-face with a stranger, and searching for a means to decipher his significance, appearing at the beginning of her journey into the troubled past of Gallipoli, the battleground for peoples with different histories, Turks, on their own soil, and Australians and New Zealanders, thousands of kilometers from home. The soldier who captures her imagination, entering the compartment of the train she had thought to hold for herself, a figure of uncertain intent, evil or angelic, this figure, this man of war, imposes a presence on her. And he imposes a narrative for her to construe. Ricoeur writes in his "From Text to Action" that "Narratives embrace someone saying something to someone about something" (p. 83): self then is inextricably connected to the other and without the other's narrative, the self cannot be a moral agent (Kearney, 2002, p. 27).

The stranger in the uniform was too thin. It was impossible not to look at him, sitting across from her in the train compartment. That he did not sleep well was evident, explicitly written on his face and body. He must have had a back injury that caused him to clench his teeth against the pain, a pain so intense it often left him doubled over even while sitting. Wounds and disease had aged him prematurely. His pain allowed his grief to be revealed, and made

---

[2]The other people represent here the people who have no face, the plurality that we cannot know (Ricoeur, 1990, p. 195).

empathy with him easy. Nevertheless, she was loathe to initiate a conversation. But her wait did not last long. He faced her, bearing the traces of intense pain, and smiled at her, just the shadow of a smile, a sign of a conversation that might begin soon (Scates et al., 2015, pp. 25–26).

To interpret the soldier through his wartime story brought to life by Bruce Scates, Rebecca Wheatley, Laura James, and yet also through the place, Gallipoli, Ricoeur's emphasis on the idea of crucial openness to the other offers this chapter a productive way forward. The moment when eyes meet across a century, the possibility of an engagement through this openness to the other: recounting history by way of exchanging stories allows the possibility of seeing each other through the eyes of the other. She may see the pain of injuries and we recognize that the interpretation of this moment must be an "ethical contact" and the narrations by Scates offers us also "...demands always striving to make the other that little less alien, [so] that we can tender (however provisionally) different interpretations of this or that other. And it is ultimately, in tune with such hermeneutics discernments, that we may offer some tentative judgments about what kinds of others we have before us" (Kearney, 2003, p. 160).

His name was Harold. Despite his wailful face and mournful voice, his story became blessed for an instant while mentioning about his love. He was about to marry his sweetheart and she was the best thing that had happened to him... His story, on the other hand, was tragic. He had shrapnel wounds that cut across his hands and body (Scates et al. 31), and his rheumatism was getting worse. The doctor had noted the gnawing pains in his leg joints, the wasting of his calf muscles and damage to his knees. Because of these he wondered if he could ever earn enough to keep a wife and perhaps a family (Scates et al. 32). Despite being buried alive by a shell, despite his shredding by shrapnel, a partial pension was the best Harold might hope for.

She was known as talkative, but this time she preferred to keep silent. Harold's story, his fragile appearance aroused a feeling in her. It was not a mercy, but a strong feeling of empathy, full of respect.

108 Language and Cross-Cultural Communication in Travel and Tourism

Listening to him through his own words lets the woman open herself to the story of the other across a century by way of Scates' narratives. This opening can be seen, in Ricoeurian terms, as a translation. The woman does not make a linguistic translation, however, but an ontological one.[3] Ontological translation, translation of experience between actors within the same linguistic community,[4] is needed since "even speaking one's own native language is already to translate" (Ricoeur, 2006, p. XV). Ontological translation provides the woman a self-understanding through his narrative since "there is no self-understanding possible without the labor of meditation through signs, symbols, narratives, and texts" (Ricoeur, 2006, p. XV). Through them, self-knowing becomes possible for the cogito as Ricoeur (1990) suggested "To say self in not to say I" (p. 180).

The woman, at that moment, forfeits her native language's claims and its nationalism in order to welcome to the other. Ricoeur (2006) calls this as "Linguistic Hospitality." For him, this requires us to renounce the omnipotence of our narratives. "It is always possible to tell the story in a different way. Linguistic Hospitability, therefore, is the act of inhabiting the word of the Other paralleled by the act of receiving the word of the Other into one's own home, one's own dwelling" (pp. 19–20). Harold's narrative about himself becomes important for the women listening to him in that regard: his narrative of structure, words, sentences through the signs and symbols he chooses, both form the ethical base of and give an opportunity to interpret Harold being an exact other. His narrative, including a common pain of humanity, loss, suffering, and love, would bring the woman and all of the others together a peaceful future.

Harold continued: The modest worker's cottage had been transformed. Wedding decorations lined the walls and the windows (Scates et al., p. 35). He seemed really excited while recounting his love. In spite of his deep grief, his tired-looking eyes were bright with desire, he forgot his pain while mentioning his love. Listening to Harold, she thought of him as a lively man who fell in love, instead of a soldier who just got out of battle. She remembered

---

[3]On Translation (2006), Ricoeur outlines two different types of translation: Linguistic and Ontological. Since there are many languages, the first paradigm involves translation between different languages. On the other hand, ontological translation, which is one of the important paradigms for this study, occurs between self and the other in the same linguistic community.

[4]Since it is a dream, the Turkish woman and the Australian soldiers speak the same language.

Transfiguring the Troubled Past Through Narratives 109

her father's stories about her grant-grandfather when she was a little girl; she had seen his letters, full of love to her grand-grandmother. Nothing different in this great emotion, love, she thought. However, the times for talking about love were limited and his memories on war were following him hard afterwards. He kept telling scraps of stories from the war.

She knew how difficult it was being involved in a battle. Her great- grandfather could escape the horrors that followed him only a few months after the war; even though, he was a strong man. He suffered so much, like Harold. They both have seen the unimaginable on the battle field. The stories Harold recounted resembled the ones her father recounted with only one single nuance: the names. Still, all stories strictly connect to humanity. Harold's best friend, who was like his childhood brother, was KIA right in front of his own eyes. He could not forget how he fell into the last place he would ever stand. Since then, Harold could not sleep properly. He knew very well before joining the war that he could be sacrificed, but he could not stand his friend lying dead.

His stories made her fill with tears. She now totally forgot the man narrating his story was a foreigner, but understood that his soul was damaged by this tragedy, and regretfully knew that his new life with his bride to be, could not possibly survive a long time. At the time of this tragic talk, the door of the compartment opened and the porter asked if they wanted to drink tea. She thought it would be better to come up for air, since she could comprehend his situation, which was full of similarities with her great-grandfather. This honorable young man and his future would not be as shiny as the words he used to predict it. Therefore, she excused herself for a few minutes and went to the door.

She recognized that except for the protagonist, there was no difference in the stories of both sides. Harold did not use any dichotomy, the only thing resonating in his story was humanity and the grief that could be shared on both sides. The stories have just become richer and deeper, and are enriched by alternative narratives. As Ricoeur (1996) suggests "The ability to recount the founding events of our history in different ways is reinforced by the exchange of cultural memories. This ability of exchange has as a touchstone the will to share symbolically and respectfully in

110 Language and Cross-Cultural Communication in Travel and Tourism

the commemoration of the founding events of other cultures..." (p. 9). However, this plurality and flexibility do not mean everything is relative. Narrative plurality does not mean lack of respect for the singularity and uniqueness of a historical event writes Kearney (2007, p. 156). On the contrary, this plurality of narratives that others recounted "... increase our sense of awareness of such singularity, especially if it is foreign to us in time, space or cultural provenance"[5] (Kearney, 2007, p. 156). Narrative flexibility allows story's to be told and retold from different perspectives. The stories she had heard were narrated by different people and different sides, both by her own family and the people on the train. Recounting of his story and her listening give the opportunity of actively resisting arrogant national narratives, which can be transformed into fixed dogma. However, narrative flexibility prevents the woman and Harold from perceiving the radical implications of nations through recounting historical events differently by different nations with empathy. She could understand the dramatic experiences and human dimension of the battle instead of seeing it as epic stories of nations.

She could not determine how long she was out of her compartment. However, realizing that the train was more crowded than she thought, wandering around the corridor made her feel better. She was just looking at the people hanging around, chatting and watching the scenery. The interesting thing was that all the people she saw were foreign and they all seemed mournful to her. After realizing the sadness of the people, she stayed there a little bit longer; however, she did not see anyone familiar. Everybody, without exception, was talking about the battle. These strangers had many different stories. Without noticing, she found herself eavesdropping on people. There was misery, grief, missing, sacrifice, expectance, and hope in these stories. Without exception. While she was home, far from these foreign people, she already had all these feelings by herself, since all the stories recounted were familiar to her. She had been narrated these stories long ago by her family. In these old stories, there was misery, grief, sacrifice, exception and hope as well. However, what was the

---

[5]According to Richard Kearney, multiple perspectives need not betray the concrete specificity of an historical event. On the contrary, they may eloquently testify to its inexhaustible richness and suggestiveness (Paul Ricoeur and the Hermeneutics of Translation, p. 156).

difference? The only differences, she thought, were protagonists: depending on which sides were narrating the war, the stories, not the emotions, changed. The war itself and the horrific things all these people had to overcome were common for each side. She recognized now that the common feelings with these strangers made this journey truly ethic. She returned to her compartment. However, as if Harold never existed, the compartment was empty. She remembered he did not have any belongings like her, but it was still strange that the man disappeared in only a few minutes. Still, his voice was resonating in the small compartment.

She was ready to welcome more strangers and listen to the stories they share of that event many years ago in Gallipoli since these encounters from the past, according to Kearney (2007), can actually give an unfulfilled future to the past (p. 157).

She did not have to wait long. Somebody knocked the door and asked for permission to enter. There was a gentleman standing in front of the door. He explained that there had been a woman in his compartment with three children and it was not possible for him to rest there. Although she was still thinking of Harold, she knows she should welcome other travelers. Her new pilgrim, in spite of his need to rest, did not seem like a quiet one himself. He gave the impression of suffering and the need to talk rather than to rest and the very minute he sat, he started to talk.

His name was Major G.F. Stevenson (Scates et al., 2015, p. 50). He had a long and touching story. He said that he had had one of the most difficult missions in the war: informing the families whose sons had been killed. The last letter he had to write was to Mrs. Lyall, which affected Stevenson deeply. The Major was careful to choose the words a grieving mother longed to hear (Scates et al., 2015, p. 52). Stevenson thought that Mrs. Lyall would read this letter time and time again. He set Lyall's personal things like his watch, some letters and a soiled pocket diary aside and sent them with his name to his mother.

The woman listening to Stevenson could understand very well how painful this would have been for him. Her own great-grandmother had had a letter from a Turkish general in 1915, informing her that her husband had been killed in Gallipoli. These were similar periods for women who were waiting for news from the battlefield, only to receive a letter of grief and reading it several times to find comfort that God had taken the life of her loved one. Here there are similar periods with different stories of a loved one's demise, with tristful letters on both sides.

The narrative of Stevenson resembled her great grandfather's in many aspects: the words they chose, the feelings they expressed, their tone of voice, all deeply humanist expressions. Simply no hatred. Someone unaware of the situation could think, listening to these men, that there was no war between their two nations. During Stevenson's story, the women may have inquired into a connotation of power about his nationalism Stevenson might have done. However, there was no masked power under the name of nationalism in neither Stevenson's nor Harold's or her great grandfather's stories, rather, only sadness in informing a mother about her son, and a humble sincerity in the willingness of communicating his difficult task.

She woke up suddenly, as if she had slept for a long time. Without reading all the stories in Bruce Scate's book, she had sunk into a deep sleep in her room with a view of the Ari Burnu memorial. An interesting dream had quickly descended upon her. She remembered having the book lying down on the sofa, but being able to read only two stories: Harold's and Stevenson's. While thinking of her family elders narrating the Gallipoli battle simultaneously, she dozed off. There was no exact answer to why she wanted to read about the Gallipoli battle, but it became a pattern for her; reading books on that battle time to time and thinking of the history of her homeland and the people who came here for battle.

She found her bookmark under her pillow and put it in her book. She checked her smart phone to see if anybody had called. It was almost 9 PM. She stood up and went to the kitchen to brew tea.

Arı Burnu is a memorial now to heal through exchanging of memories between native population in Çanakkale and people coming from hundreds of kilometers away to commemorate: a monument of remembering and mourning together for the living. Ari Burnu, as a historical memorial

Transfiguring the Troubled Past Through Narratives 113

provided the women and all the people visiting and living there inspiration to encapsulate ethics of hospitality, flexibility, plurality, and all paving the way toward a transfiguration of the past. This memorial[6] welcomes the story of the other, invites the empathy toward others by exchanging their narratives and recounts the founding event of national histories in different ways. And Arı Burnu is a home for the ones who engage with the battle in 1915 as Ataturk clearly states in his letter to Anzac[7] Mothers:

Those heroes (Anzac) that shed their blood and lost their lives ... Therefore, rest in peace. There is no difference between the Johnnies (represent Anzacs) and the Mehmets (represent Turks) to us where they lie side by side now here in this country of ours ... you, the mothers (mothers of Anzac), who sent their sons from faraway countries wipe away your tears; your sons are now lying in our bosom and are in peace. After having lost their lives on this land. They have become our sons as well.[8]

The sacred place, ArıBurnu, itself; the narrators challenging the us and them binaries and arrogant national narratives; the woman welcoming others, listening to their stories, struggling to interpret the diverse kind of otherness are in line with the ethical sense of Ricoeur and Ricoeurian interpretations of Kearney. Ricoeurian thought required a kind of narrative understanding, and provided the unnamed woman an encouragement to understand the contemporary civilization she lives in. While listening to the unfamiliar, she discovered the other in herself, and of course, herself in the other through various kinds of myths, symbols, signs, or analogies. In other words, through others, she herself self-recognizes by identifying herself with other people.

Facing with the stranger/other/unfamiliar, she fulfilled the ethical task of remembrance, she interpreted otherness by their way of talking, body language, and perspectives. Through Ricoeurian philosophy, the hermeneutic modal of memory constituted otherness without opposition

---

[6]"On Anzac Day (25 April) each year, Australia and New Zealand host the Dawn Service at the Anzac Commemorative Site on the Gallipoli Peninsula, with the gracious assistance and cooperation of Turkey" (https://turkey.embassy.gov.au/anka/GallipoliAnzac.html).

[7] ANZAC is an acronym for Australian and New Zealand Army Corps, a grouping of several divisions created early in the Great War of 1914–1918.

[8] Atatürk's letter to Anzac Mothers in 1934 at Gallipoli. This letter appears on the Atatürk Memorial, ANZAC Parade, Canberra.

to selfhood. The path the woman followed was truly ethical in terms of Ricoeurian philosophy. The acts of the people recounting their stories are explicitly ethical in terms of Ricoeurian ethics as well. Their stories are worth recounting. The integration into their life narratives brings into the open about their relations with others. Therefore, recounting their narratives articulates their point of view involving others.

6.4 CONCLUSION

Life is not only a biological phenomenon. It can be interpreted in various perspectives and Ricoeur lays out a perspective of narrative. According to him, life can be understood only through the stories as we tell about it and the examined life is the narrated life (Ricoeur, 2002, p. 31). He borrows in the sense of word "the examined life" from Socrates and rework "The unexamined life is not worth living" (Wood, 1991, p. 10). Therefore, there is a strong connection between life and narrative. In living, we create the story of our life (Simms, 2003, p. 104) and the narrative about life should involve the narrative of the other.

In this light, concluding this study ratifies the hypothesis and answers the questions put forth at the beginning. Building a hermeneutical bridge to understand the other, to have the other's narrative in the narrated stories is a way of being an ethical subject. It is explicitly comprehensible that both sides of the story, the woman who was open to the different recounted stories of the same battle; and the people who were sharing their stories with her recounted their stories through linguistic hospitability when their eyes meet across a century.

KEYWORDS

- hospitality
- narrative
- Paul Ricoeur
- hermeneutics
- other

REFERENCES

Kearney, R. Between Self and Another: Paul Ricoeur's Diacritical Hermeneutics. In: *Between Suspicion and Sympathy Paul Ricoeur's Unstable Equilirium*; Wiercinski, A., Ed.; The Hermeneutics Press, 2003; pp 149–171.

Kearney, R. Paul Ricoeur and Hermeneutics of Translation. *Res. Phenomenol.* **2007**, *37*.

Kearney, R. Strangers and Others: From Deconstruction to Hermeneutics. *Crit. Horiz.* **2002**, *3* (1), 27, 7–36.

Ricoeur, P. *From Text to Action*; North Western University Press, 1991.

Ricoeur, P. *On Translation*, Ed. Kearney, R.; Sage Publications: London, 2006.

Ricoeur, P. *Oneself as Another*, 195, Trans. Blamey, K.; The University of Chicago Press: Chicago, 1990.

Ricoeur, P. Reflections of a New Ethos for Europe. In: *Hermeneutics of Action*; Kearney, R., Ed.; Sage Publications: London, 1996.

Scates, B. et al. *World War One: A History in 100 Stories*; Penguin Random House, 2015.

Simms, K. *Paul Ricoeur*; Routledge, 2003.

Wood, D. *On Paul Ricoeur, "Life in Quest of Narrative"*; Routledge, 1991.

CHAPTER 7

Literary Ethnography and Travel Aesthetics: Amitav Ghosh's The Hungry Tide and Jungle Nama: A Story of the Sundarbans

NOBONITA RAKSHIT and RASHMI GAUR

Department of Humanities and Social Sciences, IIT Roorkee, Roorkee, India

ABSTRACT

Because of the limited production of "travel writings" in India, academic attention to its literary perspectives has been remarkably lacking. For a scholar of literary studies, this is unfortunate because travel narratives compellingly reflect the sociocultural conditions and cross-cultural practices across the world, resulting in an ethnographically sensitive and stylistically complex aesthetics. The essay will, thus, first situate Puri and Castillo's call for "theorizing fieldwork in the humanities" (2016) as an important reason behind encapsulating the lived experiences of local people as residents and the writer as a literary ethnographer in travel narratives in the postcolonial nations. Then, drawing from recent scholarships on "literary travel" that strategically adapts the fieldwork narrative and literary narrative through the regional travel aesthetics in postcolonial studies, the paper will show how the noted Indian novelist Amitav Ghosh, working across a range of geographic contexts, in his *The Hungry Tide* (2004) and *Jungle Nama: A Story of the Sundarban* (2021) historicizes the tragic fate of the people of Sundarbans and their embodied experiences in the face

---

Language and Cross-Cultural Communication in Travel and Tourism: Strategic Adaptations.
Soumya Sankar Ghosh, Debanjali Roy, Tanmoy Putatunda, & Nilanjan Ray (Eds.)
© 2025 Apple Academic Press, Inc. Co-published with CRC Press (Taylor & Francis)

of postcolonial modernity marked with capitalist–carbon economy and increasing ecological catastrophes through his extensive travel to Sundarbans as part of his literary endeavor. Closely engaging with the textual practice of intertwining local dialect and indigenous belief systems in the Western literary forms, specifically the interplay of socioecological issues with his experimental use of the myth of Bon Bibi, without compromising their cultural and performative aspects, the paper argues that Ghosh's travel aesthetics is literary–ethnographically aware and extensively conscious of the constructed and negotiated differences between the lived experiences of fieldworks and abstract knowledge of travelistic encounters.

7.1 INTRODUCTION

In 2003, Shobhana Bhattacharji, in her article lamented that "Despite regularly publish travel writings like the Himalayan Journal or travel guides, discussion of Indian travel writing in English as a genre is itself somewhat nascent. It too at the stage of requiring description and categorization ... the been-there-done-that aspect of guide books ... are unmemorable because there is little variation possible in travel writing" (p. 200). Whether nascent as a genre or requiring description and categorization, the situation has not progressed much in the last 20 years. Particularly in the context of India, the lack of production of travel writings leads to scant academic attention to its literary perspectives. However, with the use of travel as a central theme of fiction or archiving the narratives of travelistic fieldwork in non-fictional works, the postcolonial writers as custodians of cultural memory and social awareness draw renewed attention to travel narratives. What is particularly important for ethnography-inclined literary practices is that these travel narratives compellingly reflect the world's sociocultural conditions and cross-cultural practices, resulting in an ethnographically sensitive and stylistically complex aesthetics aspects the paper is reading as "literary ethnography." The essay will, thus, first situate Puri and Castillo's (2016) call for "theorizing fieldwork in the humanities" as an important reason behind encapsulating the lived experiences of local people as residents and the writer as a literary ethnographer in travel narratives in the postcolonial nations. Then, drawing from recent scholarships on "literary travel" that strategically adapts the fieldwork narrative and literary narrative through the regional travel aesthetics in postcolonial

Literary Ethnography and Travel Aesthetics

studies, the paper will show how the noted Indian novelist Amitav Ghosh, working across a range of geographic contexts, in his *The Hungry Tide* (2004) and *Jungle Nama: A Story of the Sundarban* (2021) historicizes the tragic fate of the people of Sundarbans and their embodied experiences in the face of postcolonial modernity marked with capitalist–carbon economy and increasing ecological catastrophes through his extensive travel to Sundarbans as part of his literary endeavor. Closely engaging with the textual practice of intertwining local dialect and indigenous belief systems in the Western literary forms, specifically the interplay of socioecological issues with his experimental use of the myth of Bon Bibi, without compromising their cultural and performative aspects, the paper argues that Ghosh's travel aesthetics is literary–ethnographically aware and extensively conscious of the constructed and negotiated differences between the lived experiences of fieldworks and abstract knowledge of travelistic encounters.

7.2 PURI AND CASTILLO, HUMANITIES FIELDWORK, AND TRAVEL FICTIONS IN INDIA

In 2016, Shalini Puri and Debra A. Castillo wrote, often experimentally, the role of fieldwork in humanities in general and literary studies in particular in capturing a community's lived experiences in the practices of literature and literary criticism. They claim that literary studies, along with philosophy, is one of the humanities disciplines where fieldwork has been least common (p. 12). The dominant images and well-known paradigms for literary scholarship such as the silent archives, white-marble tower, cultured metropolis, etc., distance the researcher from the chaos of everyday life (p. 12). While the fieldwork is neither a standardized, prescriptive, or paradigmatic approach to perceive everyday life, nor the sole path to knowledge, nor with any imprint of authenticity, it offers various archives, means of encounter, engagement strategies, and interlocutors and therefore, possibly varied insights. It can expand our understanding of human and the non-human environment in such a way that also requires an estrangement and reevaluation of the self. This is the "homework" that makes the fieldwork feasible (p. 17).

While this concept of "homework" could be applied as well in the case of fieldwork in Anthropology, a discipline that converses frequently with a diverse range of people rather than only specialists and elites, what is clear is that fieldwork in literary practices changes the way it interacts with diversity from text to day-to-day existence, initiating a movement from the boundaries of text and the limitless nature of existence unbounded (pp. 5–9). An anthropologist fieldwork having unavoidable historical connections to colonial enterprises, holistic cultural models, exoticized treatment of distant others in confined cultures, the recurrent misinterpretation of indigenous knowledge, integration into empiricist philosophical theory, and its disadvantages and authority inherent in the model of ethnographer (p. 7) would never have enabled gaining knowledge via experience from below (Spivek, 2005, p. 36) that would be key to theorizing fieldwork in literary practices in a postcolonial context. Its extensive conversation with a broad section of people would be a necessary but insufficient condition for analyzing one's cultural values and epistemology (Puri and Castillo, 2016, p. 9). Hence, Puri and Castillo (2016) stress the inseparability of literary text and the experience in fieldwork to draw attention to biases, hints of glimpsed and implied realities beyond academic vision, the requirement of labor, the difficulty of using involved lenses, and consideration and analysis of parallels and differences (p. 8). For them, Anand Pandian (2015) notes that experience involves experimenting with life which becomes an arena of hypothesis, testing, and challenging lessons. Then, the fieldwork, for them, becomes not simply travel or presence but also thinking, reflecting, analyzing, undertaking, and trialing and moreover crossing the boundary between real-world experience and theoretical possibilities (p. 11).

With the rise of fieldwork as a sustainable practice in humanities, Puri and Castillo's idea of humanities fieldwork as, in Pandian's (2015) words, an opportunity to challenge and involve the unfinished, open-ended characteristics of life, to observe things as they unfold, to include the ambiguity and vulnerability of human–non-human relations into the core of our intellectual labor gains widespread attention. With the intellectual exhilaration, interdisciplinary approach, intersubjective connection, and professional comprehension in academia, that attempt to position fieldwork in humanities as a counter to the "erasure of the global South" (Puri and Castillo, 2016, p. 12) and Puri and Castillo's (2016) call for launching a public discussion among scholars and academicians conducting fieldwork in the disciplines of humanities to increase awareness of such methods without attempting to standardize them prove pivotal for the emergence of travel-documentation projects in postcolonial nations. However, documenting the narratives of travel in India is not a sudden phenomenon. Shobhana Bhattacharji (2003), working on articulating, shaping, and reshaping the genre of Indian travel writing in English, reminds us of the vast range of travel writings available—"general, specialized, amature, bureaucratic, religious, secular, endless school and college essays on journeys by train and bus, letters home by travellers" (p. 199). However, reading the erasure of travel narratives of postcolonial nations from the mainstream cultural imagination, Bhattacharji (2003) recounts that the travelers who write about India are white Europeans, mountaineers, and trekkers for whom the travel is not fieldwork but a medium of recreation. She notes the "vast number of refugee travellers in India—migrant labour, economic and political refugee from one region to another—usually do not write about their experience. At most they write brief letters, or a line on a money order, or get someone else to write for them, and if they do, it is not likely to be in English" (p. 199). Travel-documentation project, which archives both the background narratives of a journey taken as a medium of embodied encounter and travel as a narrative mode, could then be seen as providing the much-needed succor to Puri and Castillo's call for literary ethnography. Therefore, in the postcolonial nations, an alternative understanding of travel writing emerges both as breaking the traditional boundaries of disciplines like anthropology, history, and fiction (Dixon, 1996) and fending off the influence of urban supremacy and globalization through recording embodied experiences which are frequently dissociated from predominantly European or North American construction of ontologies (Edward and Graulund, 2010, p. 2). This focus on the everyday life of ordinary people and the embodied experience of the travel writer that counters the abstract knowledge of a place, people, and environment that appear predominantly in contemporary postcolonial travel narratives. Edwards and Graulund (2010) tell us that such writings begin to appear through postcolonial writers such as Amitav Ghosh, Pico Iyer, V. S. Naipaul, Jamaica Kincaid, Jan Morris, Charlotte Williams, and others who attempt to engender among readers a growing sense of decentering the predominant systems of neo-colonization through the depictions of travel (p. 3). Their innovative aesthetics treatments of travel in fiction are vital in developing the genre in India.

The literary fiction that uses travel to encapsulate ordinary people's lived embodied experiences is not peculiar to postcolonial nations. This is so because Maria Laurdes (2003), writing about travel writing and postcoloniality, observes that the aesthetics of travel can no longer be used as "an instrument of colonial expansion." Instead, it has become an effective tool for cultural criticism (p. 51). Laurdes (2003) notes that rather than delighting the readers through the representation of exotic or bragging about nationalism, counter-travel writing seeks to unsettle the reader's contentment by unmapping their preconceived worldviews (1997, p. 143). While doing this, the counter-travel writings dismantle the Eurocentric viewpoints from where this genre originated (52). According to Justin D. Edwards (2010), by articulating a sociopolitical narrative that represents postcolonial experiences from postcolonial nations, postcolonial travel writings offer the potential for cultural critique (p. 105). As literary critic, Upasana Dutta (2013), working on the modern direction of travel writings in India, further informs that Indian travel fictions hold more potential and well-structured literariness than it did before. Since fiction writers are no longer required to document a location's exterior details superficially, they are free to explore the nuances of a particular community and their sociocultural practices. This is similar to how, with the invention of photography, painting could focus on attempting to encapsulate the soul of things rather than just a strict surface representation of a subject (p. 77).

Here, it is crucial to note that postcolonial travel fiction uses travel as an aesthetics medium and a method of research, which allows it to address both the conflict of belonging and broader sociohistorical issues (Edwards, 2010, p. 106). As Edwards argues, such fictions are deeply impacted by subject orientation, social awareness, cross-culturality, and cultural syncretism. For Edwards and Graulund (2010), these impacts bring forth several aesthetics changes, including the critique of capitalism's cultural production and its current transformations, merging the material circumstances of the locations the writer visits with the ideas of the Empire (p. 8), questioning the concepts of genuine and truthful travel by asking: Whose authenticity we are seeking in a travel writing? (p. 9) and finally postulating travel as a continuous renegotiation of several, equally legitimate forms of transportation; an activity which is neither worse nor better than it was 100, 1000, or 2000 years ago. It is only unique and different (p. 13). Therefore, it can be argued that Indian travel fiction not only focuses on the linear narrative of the physical journey but also focuses on the metaphorical journey where the readers enter into the narrative space and travel with the author in the to-be of the future-to-be—and which the chapter would read now as "literary travel."

Literary Ethnography and Travel Aesthetics 123

7.3 TRAVEL AESTHETICS AND LITERARY TRAVEL IN TRAVEL FICTIONS

Regarding the travel fictions mentioned above, Tabish Khair (2005) suggests that they focus on non-European travelers whose travel experiences have been erased in two ways: (1) erased because they left few or no written records and (2) erased by writing a narrative based on knowledge formulated by gypsies, enslaved people, lascars, etc.; and explore the politics of movement to understand why the non-European travelers' movements narratives are undocumented despite their reference in the texts of European travelers (pp. 5–7). In these travel fictions, there are now themes like the encroachment of globalizing forces into local, the changing processes of knowledge production, the ecosocial exploitation, the experiences of forced displacement and voluntary dwelling-in-travel, the histories of border-crossing, diaspora, and migration, among others— exploring the sociopolitical ecology of the global South by the experimental use of fieldwork experiences in literary narratives. In these novels, as William Dalrymple in conversation with Tabish Khair (2010) observes, the compassionate traveler who immerses himself/herself in a country may acquire not simply factual information. Rather, the traveler will gain an aesthetics as well as emotional perception regarding the human psyche and their multidimensional responses towards various events and objects which might never be obtained via library reading (p. 184). According to Dalrymple, travel writing can offer glimpses into daily life because the commonality of people's lives is rarely portrayed in academic writing and is rarely brought forth by any other discipline. Despite the technological advancement and the communication revolutions, quality travel writing is still incomparable (p. 184). A representation of these abstract and yet embodied experiences of the author–traveler as a literary ethnographer creates an aesthetics challenge because a writer (whose journeys are often considered as privilege) needs to be careful to write about the blanks— "insufficient stories, insufficient knowledge about a place or event, stories that have not been corroborated by other stories, personal stories that have not been corroborated by newspaper stories or "facts" (Bhattacharji, 2003, p. 205).

Writing about the interrelation of the writer's experience of travelistic fieldwork and the literary-ethnographic production, Naipaul, in a conversation with Ahmed Rashid, observes that his books can be called travel writings not in the sense of route description, but as a work of fiction where he travels to enquire. His writing is not that of a journalist but rather of a creative writer who combines inquiries with empathy, awareness, and curiosity. His novels are primarily built narratives in which there are various minor storylines that are interconnected in a broader network in addition to the inquiries (Naipaul, 1996, p. 16 as quoted in Borm, 2004, p. 22). Such a framework allows us to perceive how travel as an inevitable part of the writing process has developed a triad between the traveler–author, the resident protagonists, and the readers of the literary ethnography. Naipaul's concept of "travel writing" then encourages us to engage in a reading of travel less as a metaphor for a journey within a "physical space than in the space of a narrative" (Jean-Didier Urbain as quoted in Borm, 2004, p. 23). Such a concept of literary travel that conflates the physical fieldwork narrative and the metaphorical–literary narrative through the regional travel aesthetics becomes key in exploring the postcolonial travel fictions.

The cross-cultural, socioecological, and economic–political changes that have defined and redefined the experiences of travel in the postcolonial countries, as mentioned above, compelled a travel-aesthetics (a novelistic form) that had to balance traditional disciplinary stylistics and contemporary cross-border approaches. Now, if the postcolonial travel writings have already used an aesthetics that subverts the "colonial travel narratives" and the use of travel as a "tool of empire" (Edwards and Graulund, 2010, p. 3), the Indian travel fictions demand further regional travel aesthetics to focus on the undocumented everyday lives and their miserable conditions of displacement, disjunction, and rupture. The Indian novelists adopt various regional cultural art forms that they encounter during fieldwork, such as epics, nautanki, local myths, Jatra, etc. implement them into their narratives to provide a glimpse of ontologies existing outside the borders of European knowledge production. Thus, in the complex narrative pattern, both the experiences of travelistic fieldwork and literary narratives predominantly coexist, and the task of the narrator is not to exoticize or foreshadow the cultural nuances but to bring it into the fiction to remind us how the postcolonial communities, marked with exploitations of the capitalist–carbon economy and increasing ecological catastrophes, survive and resist these exploitations. This travel aesthetics may be read as "literary travel" as it not merely excavates the fieldwork experiences of the writer–traveler but also brings to the cognition the marginalized cultural forms long suppressed

Literary Ethnography and Travel Aesthetics

in the mainstream cultural imagination. Nonetheless, travel as a mode of representing the embodied–emplaced experiences does not mean to reproduce the actual events or occurrences literally into the narrative structure but to interrupt the straightforward process as merely given. When questioned by Pooja Bhula about whether his travel experiences inspire his stories, Amitav Ghosh replies that while writing his first novel, all his experiences gained as a researcher of social anthropology, lessons of Arabic, fieldwork in Algerian Sahara, Spain, and Morocco find their ways into the book (Ghosh, 2019). As the chapter will argue, through a close reading of Amitav Ghosh's *The Hungry Tide* and *Jungle Nama: A Story of the Sundarbans* to represent the importance of travelistic fieldwork in exploring the embodied experiences of postcolonial modernity, the author needs to mediate between the distant, abstract knowledge and local, embodied experiences. In it, the conventional definition of travel writing may be redefined. Ghosh's travel aesthetics is literary–ethnography oriented not only because it derives from a confluence of fieldwork narratives with literary narratives but also because it is extensively conscious of the constructed and negotiated differences between the lived experiences of fieldwork and abstract knowledge of travelistic encounters. In what follows, this chapter will first briefly situate the postcolonial dimensions of these two narratives (suggesting Ghosh's fieldwork-based writing tendencies) and then show, through his use of myth and history, how Ghosh produces a complexly framed literary travel aesthetics.

7.4 *THE HUNGRY TIDE* AND *JUNGLE NAMA*: A LITERARY ETHNOGRAPHIC APPROACH

From the beginning of his writing career, Ghosh has widely captured his travel experiences as fieldwork both in fiction and non-fictional works. *The Hungry Tide* and *Jungle Nama* follow similar premises. *The Hungry Tide* tells the tale of the people of Sundarbans whose lives are heavily impacted by ecological exploitation, human trafficking, encroachment of destructive modernizing processes, and sociopolitical violence that results in a genocide, namely the Marichjhapi Massacre. Amitav Ghosh, like a literary ethnographer, first takes the urban-modern readers into the tide country of Sundarbans through the characters of Kanai and Piya and then makes them face the undocumented history of a massacre and rural–regional cultural arts and lifeworld through the character of Fokir and Nirmal.

At the heart of *The Hungry Tide* is a clash between abstract knowledge and embodied–emplaced cultural encounter. Embodiedness comes from embeddedness into the tide county where the new lands are created overnight, covered by leathery mangroves and impassable foliage. The landscape is both utterly hostile and resourceful to human beings and determines to either destroy or expel them (Ghosh, 2004, p. 8). Fokir, a boatman "who'd grown old on the water" (p. 46), is well aware of the strategies of survival and cultural traditions of watery world of Sundarbans. He believes in the tales recounted by his mother, Kusum, about the protective nature of forest goddess Bon Bibi, the role of whales as Bon Bibi's messengers of apocalyptic events, the everyday rules of survival that are being followed from generations, the ceremonies to be performed to escape the tigers in the forest and water. Thickly covered with mangroves, perishing thousands of people in the impenetrably thick foliage, attacked by royal Bengal tigers, poisonous snakes, and dreadful crocodiles (p. 8), the utterly transformed world of Sundarbans (a land which is "half-submerged at high tide" and the forest only emerges in low tide, hence tide country) has developed its own tale. In a place that is extremely prone to cyclones, floods, sea-level rise, illegal fishing, and utter poverty, life is full of economic hardships, and yet people celebrate the traditional ceremonies, enjoy the performance of the Jatra, catch fishes, collects firewood, and chants the myth of Bon Bibi and Shah Jangoli throughout the year. Most of these ceremonies are documented both orally by the inhabitants and in written form by the people who traveled to the tide country to, at first, conquer and then for research purposes. The novel begins with such an expedition taken up by Piya to "do a survey of the marine mammals of the Sundarbans" (p. 11) and her encounter with Kanai travelling to Sundarbans in order to explore his uncle Nirmal's letters to him. The travels would turn out to be an exploration into the histories of colonial invasion, rural-cultural vibrancy, sociopolitical exploitation, and the ever-changing ecology of Sundarbans. The story subsequently explores how Ghosh's travel and engagement with the ecosocial conditions and histories of people, place, and environment takes the readers into the 1970s Sundarbans and connects it with the current social, political, and ecological conditions.

These descriptions of the place and its impact on the lives of people would have seemed partial, and half-glimpsed realities had Ghosh not added the "author's note" in *The Hungry Tide*, where he notes that it is not only the characters but the two main settings of the novel, namely Lusibari and

Garjontola are also fictitious. However, islands such as Morichjhapi, Gosaba, Emilybari, Canning, and Satjelia, do exist and are included in the novel as has been experienced by Ghosh. The High school in Gosaba, namely, Rural Reconstruction Institute founded by Sir Hamilton is there and his uncle, Shri Chandra Ghosh worked as a headmaster of that school for more than a decade. His first memories of the tidal country are greatly owed to him and his son, Ghosh's cousin Subroto Ghosh (Ghosh, 2004, p. 428).

His repeated visits to tide country to engage in a historical and cultural reading of the place, participating in traditional ceremonies, conversation with the local folks, accompanying the cetologists like Isabel Beasley in order to study the migratory patterns of Irrawaddy dolphins, and his in-depth research on Morichjhapi massacre results into his publication of *The Hungry Tide* in 2004. During his fieldwork for *The Hungry Tide*, Ghosh comes to know of the Bengali translation of the Arabic text Bon Bibi Johuranama which he later retells in *Jungle Nama: A Story of the Sundarbans*. In the "Afterword" of *Jungle Nama*, Ghosh postulates,

"My novel *The Hungry Tide* is set in the Sundarban. The Bon Bibi legend enters and informs the novel in many different ways: in the chapter entitled 'Memory,' for instance, I wrote about some of the story's context and background...Jungle Nama is not intended to be a definitive version of the narrative; it is, rather, yet another re-telling of a story that already exists in many iteration" (Ghosh, 2021, pp. 75–76).

Moreover, the narrative construction of Sundarbans, its dynamic landscape, its cultural beliefs, and its lifeworld in his fiction is inspired by the field notes taken on a 2002 visit to the place. In his Berlin Lecture series, published in 2016 as *The Great Derangement*, Ghosh provides fragments of the field notes taken during these visits. He notes that the demonstrably alive landscape is more than a backdrop for telling human history. The land is, in fact, a protagonist. Again, in another note, he writes that in Sundarbans, even a little youngster while recounting the story of his/her grandmother will start: "in those days the river wasn't here and the village was not where it is" (Ghosh, 2016, pp. 7–8). It is this process that amalgamates these literary texts, their sense of the local, the field experiences in Sundarbans, and Ghosh's own experiences of unprecedented climatological catastrophes is what this chapter calls a "literary fieldwork." His positions as a literary writer, social anthropologist, climate activist, and academician broaden the scope of this literary fieldwork. Ghosh reconciles what he touches in

128 Language and Cross-Cultural Communication in Travel and Tourism

the field—other human and more-than-human-world—with what he writes for a living–literary fiction.

7.5 CONCLUSION: FRAMING A LITERARY TRAVEL AESTHETICS

We may note from the analysis above that Ghosh's is a fieldwork-oriented writing approach. In the "Forward" to Tabish Khair's magnum opus published in 2005, Ghosh observes that in travel writing, there is an acknowledgment that what is obvious to him may not be for everyone else. This recognition requires an acceptance of the witness's knowledge of gaps and recognition of the witness's knowingness. He further notes that because these travelers feel obligated to document what they see and hear, so many details that seem unimportant find their way into these narratives. Their tales are not organized according to teleology of race or civilizational growth, nor do they assume a universalist organization of reality. To Ghosh, what distinguishes these tales from travel writing that is motivated by ideas of "discovery" and "adventure" is their open-ness and feeling of awe (p. ix). Therefore, using the regional aesthetics encountered during travelistic fieldwork as a medium of exploring the sociocultural, -political, and -ecological aspects of a community func-tions as an enabling factor in explaining the migratory pattern of the people who are denied of having any being in the mainstream academia. It also explores the mythological consciousness of these migrants, which runs parallel to their historical consciousness, and attempts to rewrite an alternative history for these repressed figures. In doing so, Ghosh (2005) is "not seized by a compulsion to fit … into familiar narratives" of travel. Rather, his meticulousness lies in his "noting details" and "noticing the unfamiliar" (p. xii).

The myth of Bon Bibi that Ghosh encounters on one of his trips to Sundarbans also offers a platform for explaining the "planetary crisis" that has challenged a wide range of accepted assumptions and norms, including many that are concerned with literature and literary genres (Ghosh, 2021, p. 77). Kanai, who suppresses a "snort of laughter" listening to the story of this myth, explains to Piya later that this myth carries a history that is not only of Fokir but of the tide country. Translating to Piya the meaning of Fokir's song, Kanai recounts that in Nirmal's handbook, Nirmal writes about an occasion where Fokir, in his early childhood, memorizes and recited the verse concerning a tide country legend, namely the Bon Bibi,

Literary Ethnography and Travel Aesthetics 129

popularly known as the protector of the forest. What amazes Nirmal is that Fokir could neither read nor write. However, Nirmal also understood that for this youngster, those words represented far more than just a piece of folklore—they represented the narrative that gave life to this land (Ghosh, 2004, p. 379).

This textual mechanism of shifting the medium of fieldwork encounter from text to everyday life and vice-versa forges a theoretical and methodological intervention into the limitations of fieldwork and literary studies. Ghosh navigates the cultural space, the literary life of Sundarbans, and the multi-lingual aspects of the cultural texts and recounts, "The vocabulary of the Bon Bibi Johuranama is extraordinarily hybrid, being highly influenced by Persian and Quranic Arabic" (Ghosh, 2021, p. 76). From his non-fictional works, interviews, and newspaper essays, it becomes evident that his life as a social worker, activist, on-field researcher, and literary writer allows him to encounter many cultural practices in the peripheral world that he seeks to document in the written form without compromising their literary or performative aspects. His fieldwork-based works enable Ghosh to document these oral cultural art forms that are historically and esthetically significant for the narrative. Therefore, such a mechanism of juxtaposing travel narratives, a primarily non-fictional genre, with the work of literary fiction not only allows Ghosh to develop an innovatively literary travel aesthetics but also paves the pathway for fiction writers as well as readers to transcend the hierarchy of genre division immensely prevalent in existing academia.

KEYWORDS

- **travel writings**
- **literary fieldwork**
- **literary travel**
- **cross-cultural communication**
- **Sundarbans**

REFERENCES

Bhattacharji, S. Amitav Ghosh's Travel Writing: "In an Antique Land, Dancing in Cambodia" and "The Imam and the Indian". *Indian Literat.* **2003,** *47* (6), 197–213. http://www.jstor.org/stable/23341083

Borm, J. Defining Travel: On the Travel Book, Travel Writing and Terminology. In: *Perspectives on Travel Writing*; Hooper, G., Youngs, T., Eds.; Routledge: London, 2004; pp. 13–26.

Edwards, J. D.; Graulund, R. Introduction: Reading Postcolonial Travel Writing. In: *Postcolonial Travel Writing: Critical Explorations*; Edwards, J., Graulund, R., Eds.; Palgrave Macmillan, 2010; pp 1–16.

Dixon, R. "Travelling in the West": The Writing of Amitav Ghosh. *J. Commonwealth Literat.* **1996,** *31* (1), 3–24. https://doi.org/10.1177%2F002198949603100102

Dutta, U. Modern Directions in Travel Writing: Amitav Ghosh's in an Antique Land and William Dalrymple's Nine Lives: In Search of the Sacred in Modern India. *Coldnoon (Int. J. Travel Writing Travell. Cult.)* **2013,** *2* (3), 69–78. https://coldnoon.com/wp-content/uploads/2015/09/Upasana_Dutta_Jul13.pdf

Edwards, J.; Edwards, J. D.; Graulund, R. 'Between Somewhere and Elsewhere': Sugar, Slate and Postcolonial Travel Writing. In *Postcolonial Travel Writing: Critical Explorations*; Edwards, J. D., Graulund, R., Eds.; Palgrave, 2010; pp 104–115.

Ghosh, A. *The Hungry Tide*; Penguin Books: India, 2004.

Ghosh, A. *The Great Derangement: Climate Change and the Unthinkable*; Penguin: India, 2016.

Ghosh, A. *Jungle Nama: A Story of the Sundarban*; Fourth Estate, India, 2021.

Ghosh, A. 'Inside Amitav Ghosh's Travel Journal' [interview with Pooja Bhula], *Traveller* (National Geographic), Nov 23, 2019. https://natgeotraveller.in/inside-amitav-ghoshs-travel-journal/

Khair, T. An Interview with William Dalrymple and Pankaj Mishra. In: *Postcolonial Travel Writing*; Edwards, J. D., Graulund, R., Eds.; Palgrave Macmillan: London, 2010; pp 173–184.

Khair, T.; Leer, M.; Edwards, J. D.; Ziadeh, H., Eds. *Other Routes: 1500 Years of African and Asian Travel Writing*; Indiana University Press, 2005.

Pandian, A. *Reel World: An Anthropology of Creation*; Duke University Press: Durham, 2015.

Puri, S.; Castillo, D. A., Eds. *Theorizing Fieldwork in the Humanities: Methods, Reflections, and Approaches to the Global South*; Springer: New York, 2016.

Ropero, M. L. L. Travel Writing and Postcoloniality: Caryl Phillips's "The Atlantic Sound". *Atlantis* **2003,** 51–62. https://www.jstor.org/stable/41055094

Spivak, G. C. *Death of a Discipline*; Columbia University Press: New York, 2003.

CHAPTER 8

Journeying Through the Lesser-Known Indian Spaces: A Reading of Bishwanath Ghosh's Chai, Chai

PAMELA PATI ànd I WATITULA LONGKUMER

¹Doctoral Scholar, Discipline of English, IITRAM Ahmedabad, India

²Assistant Professor, Discipline of English, IITRAM Ahmedabad, India

ABSTRACT

Traveling for most Indians is often associated with popular tourist destinations. These destinations, however, reflect a very selective representation of an otherwise diverse country like India. From a traveler's perspective, India is often equated with the most quoted places (for example, Taj Mahal). Our argument in this essay is that, although these places define the rich cultural diversity and heritage of India, they fail to represent the country in its entirety. Interestingly also, the current conversation on traveling within India is garnering a new dimension, particularly, after the sudden emergence of the global pandemic. There is a new surge among travelers to explore the smaller pockets of the country which are often overshadowed by the popular tourist destinations. Bishwanath Ghosh's novel *Chai, Chai: Travels in Places Where You Stop But Never Get Off* (2009) precisely captures this version of India as he takes the readers through the smaller towns. Ghosh in the book looks at the popular Indian railway junctions and attempts to explore the junctions beyond just a connecting station. The novel *Chai, Chai* establishes an important connection with the readers in two ways: (1) by representing an inclusive geography and (2) by exploring unexplored Indian spaces.

Language and Cross-Cultural Communication in Travel and Tourism: Strategic Adaptations.
Soumya Sankar Ghosh, Debanjali Roy, Tanmoy Putatunda, & Nilanjan Ray (Eds.)
© 2025 Apple Academic Press, Inc. Co-published with CRC Press (Taylor & Francis)

"India has been the most varied and unique country I have ever visited. India feels like 10 countries or more rolled into 1" (Shiels, 2016).

8.1 INTRODUCTION

The above statement rightly comments on the geographical, cultural, and linguistic diversity of India. Sommer Shiels' experiences of traveling across the globe has allowed her to see, understand, and confirm the distinctive quality of the places in India, each geography offering an experience that is different from the other. While this mesmerized observation of the huge canvas of diversity is through the eye of an outsider, similar observation echoes among people (travelers and travel writers) from within the country. Rajat Ubhaykar, a contemporary Indian travel writer, contributes to this observation in the prologue of his book *Truck de India! A Hitchhiker's Guide to Hindustan* as he says:

India is bigger than the boundaries of my imagination, or anyone's, for that matter. You didn't have to go to the scale of the cosmos to imagine something vast-India was enough. Even as a child, it made me aware of the insignificance of my own little life, when confronted by the sight of India's multitudes, its green fields skirting the highway, the pools of mirage water shimmering on the distant hot tar. I had fallen for India, heart, mind, and soul. (2019, pp 7–8)

The current conversation on traveling within India is garnering a new dimension, particularly, after the sudden emergence of the global pandemic. There is a new surge among travelers to explore the smaller pockets of the country which are often overshadowed by the popular tourist destinations. While this shift takes place among travelers, scholars, critics, and writers in the recent past have recognized the need to (1) re-locate and re-imagine our understanding of traveling and (2) write about travel from within rather than being written on. Contemporary Indian travel writers and scholars such as Bishwanath Ghosh, Rajat Ubhaykar, Pankaj Mishra, Nabaneeta Dev Sen, among others, are changing the narrative. Their works meander into exploring the local geographies and cultures and, interestingly and rightly so, some of them are reversing the colonial gaze in writing about their experiences abroad.

Our paper stems from a shift in travel perspectives we see, at least, among contemporary travellers as well as travel writers. Alongside this shift, we also look at the idea of representation of a familiar geography from an insider's perspective.

8.2 RE-ROUTING THE COLONIAL TRAVEL NARRATIVES

Even though travel writing is essentially different from fiction and it is based more on "what the eye sees" (Theroux, 1975, p 379), the authenticity of that factual representation can always be contested. Especially with the early travel accounts on India written by the Western authors, a tendency to foreground the influence of Ethnocentrism[1] was noticed. The early travel accounts on India written by the western authors presented a distorted version of the East in general and at the same this negative representation helped them to establish their race as the most superior one as well as to make them the eligible saviours of the inferior races. In short, Ethnocentrism, as defined by Satapathy (2012), is "an unbearably selfish discourse" (Idea of the West section, para.1). In an article published in *The Hindu*, Radhika Santhanam goes one step further and discusses the severe consequences of Ethnocentrism. She states, the concept as understood by the west "could lead to prejudice, dislike, dominance, conflict, instability of democratic institutions, and even war" (Santhanam, 2022, "Ethnocentrism", para 1). Pramod K Nayar further talks about the aspect of "dominance" that directly generates from an ethnocentric attitude.

Nayar's book *Colonial Voices: The Discourses of Empire* (Nayar, 2012) attempts to trace the ruling period of the British in India until its Independence in 1947 and examines the traces of colonialism that continues to linger. Nayar's book is significant as it discusses the effects of travel narratives by the colonials into the exotic Eastern lands. Much before Indian travel authors began to write and document their travel experiences, most of the travel books based on India were written by the British scholars and administrators. While some of the scholarly works are important contributions such as that of Bill Aitken and William Dalrymple, the early writings

---

[1]Amrita Satapathy in her essay "Reconsidering the West in Early Autobiographies and Travel Writing in Indian Writing in English" defines ethnocentrism as "a form of cultural bias. Coined first by William Graham Sumner, a social evolutionist, ethnocentricity is the tendency to look at the world primarily from the perspective of one's own culture."

134 Language and Cross-Cultural Communication in Travel and Tourism

on India by the British administrators are often reductive definitions of the places and the people.

In his attempt to define and situate the observations on India by the European traveler, Nayar highlights the important technique of "narrative possession" (2012, p 54) as a part of the process of colonial project. Nayar states:

The "proto-colonial" discourse of discovery moves from imagining of what could be discovered in the East to the ordering of what was discovered. These writings therefore mark a narrative possession—we could think of it as "colonization"—of India. (2012, p 54)

This process of colonial project is further elaborated in another essay "Marvellous Excesses: English Travel Writing in India, 1608–1727" by Pramod K Nayar in which he systematically analyses the works written by the early English travelers in India. Thomas Mun's *A Discourse of Trade to the East- Indies* (1621), Edward Terry's *A Voyage to East India* (1655), Thomas Bowrey's *A Geographical Account of Countries Round the Bay of Bengal 1669 to 1679* (1905), and Alexander Hamilton's *A New Account to the East Indies* (1727) (as cited in Nayar, 2005)—are some of the significant works decoded in this context. Even though Nayar restrains from bracketing the East India Company (EIC) travelers of 1600–1750 as "colonial," he simultaneously believes that "the travelogues of this period embody themes that anticipate and prepare for the overt colonialist writings of the post-Plassey (1757) phase" (2005, p 214). Having studied the works of the early English travelers, Nayar arrives at three observations— (1) description of a landscape of plenty (2005, p 217) and a simultaneous sense of wonder and awe, (2) continuity of that "excess" with "a note of censure" (2005, p. 225), and, finally, (3) transformation of the valuable native "icons" into "grotesque exhibitionism" (2005, p 230).

The first phase that Nayar describes, projects the western gaze on the Indian land as a territory that deserves further inquiry and even monitoring if need be and reveals a tendency to exoticize. The travelers of the first phase were travelers who paid attention to "specific material/commercial features of particular places" (2005, p 218). Their travelogues used the "trope of intensification" (2005, p 219), which aroused a sense of wonder and awe. The early English travelers were particularly interested in India's productivity—in its "loads of crops, the dense woods, and markets with a variety of fruits and vegetables..." (2005, p 220), and their imagination of the productivity of the soil was further extended even to the natural fertility possessed by the Indian women as well. Very interestingly, this phase excluded the farming laborers and established the argument: "The wealth of the Indians was thus undeserved because they did not toil for it" (2005, p 222). This, rightly so, reflects Edward Said's question of representation of the Orient and its linkage with ideas of "Oriental despotism, Oriental sensuality, Oriental modes of production, and Oriental splendour" (2001, p 47) in his seminal book *Orientalism* (Said, 2001). Nayar's observation also reminds us of Said's preliminary ideas on Orientalism. Orient, Said argued, was "almost a European invention," and had been since antiquity "a place of romance, exotic beings, haunting memories and landscapes, remarkable experiences" (2001, p 47).

The second phase English travelers shifted their attitude drastically and projected India as a wild land of chaos, disease, brutality, and ignorance. The travelers started recording their discomfort with the Indian crowd and atmosphere. The third phase included travel writers who scrutinized every "excess" presented by their immediate predecessors and portrayed India as a land devoid of moral values. What Nayar concludes in his paper is that the colonial writing revolves around a pattern—a pattern that starts with an overwhelming appreciation of the colonized land and ends with the projection of a significant lack "of knowledge, kindness, rights, and freedom" (2005, p 230) in it.

Thus, the portrayal of a colonized land by the colonists has always more to do with the power politics; and less to with an innocent authentic representation. It has never been free of the controlling attitude of the rulers. While Nayar reveals a colonial pattern in their ways of Eastern representation, Amartya Sen categorizes it into three separate yet intertwined approaches: (1) the exoticist approach, which echoes Nayar's idea of the colonized land as a "landscape of plenty" (2005, p 217), and (2) the magisterial approach that reflects Nayar's idea of India as an "object of inquiry and control" (2005, p 215). The third approach deviates a bit from Nayar and refers to the curatory approach, an approach that "attempts at noting, classifying and exhibiting diverse aspects of Indian culture" (as cited in Islam and Das, 2017). Interestingly, Sen argues that it is quite possible for a British ruler to internalize all the three approaches.

Therefore, even if the early English travelogues contributed a lot to the literary tradition, one cannot overlook the underlying political perspectives. Along with their implied colonial perspectives, there was a parallel flow of explicit negativity toward India. In this regard, Nayar quotes

Thomas Herbert, who, while describing the woods and groves of trees in India, states, "These negroes you see have no famine of nature's gifts and blessings" (as cited in Nayar, 2005, p 220). Rita Banerjee in her book *India in Early Modern English Travel Writings* makes a similar argument in which she says, "Seventeenth-century Western representations of India often show an attempt to negativize India by placing her in opposition to Europe. This is a major trend, although not the only one" (2021, p 239).

On the other hand, one cannot overlook the early travel writings written by the Indians which are characterized by a kind of overwhelming appreciation and a sense of wonder for the West. This is particularly evident in the 18th-century travel narratives and up until mid-20th-century writings. Such enchanted representation perhaps attributes to the awe and wonder for the West and at the same time marks the lack of an insider's critical observation of his own country. An example is Trailokyanath Mukhopadhyay's *A Visit to Europe* in which he critiques the Indian mentality and applauds the West for its enthusiasm in new knowledge. Mukhopadhyay states:

> The difference between the European and the Indian is very marked on this point. The former is always on the lookout for new things. He is constantly trying to make new contrivances and discover new ways...Not so with the Indian. He will not accept any new knowledge even if it is forced down his throat with a hydraulic hammer. (Mukherjee, 1889)

This fascination had an indirect influence on the target readers assumed by the early Indian English writers as well. We see this in Meenakshi Mukherjee's book *The Perishable Empire* (Mukherjee, 2000). In the essay "Nation, Novel, Language," Mukherjee observes the tendency of the early Indian English writers to portray a "broad Indian identity" (2000, p 15) rather than to capture the minute regional details in their works. Mukherjee explains this with the help of Susie Tharu's concept of "addresser's discursive relationship about the addressee" (2000, p. 14) and points to the works of Lal Behari Day—*Govinda Samanta* (1874) and A. Madhavaiah- *Satyananda* (1909). The reception of Western literature in 19th-century India led to the appearance of Indian English writers who in writing assumed its readers "to be situated outside the culture, possibly in England" (2000, p. 14) and hence, the language English ostracized the regional subtlety in the emerging Indian English Literature. Mukherjee's essay is an important

Journeying Through the Lesser-Known Indian Spaces 137

reference here as it refers to the new experience that overwhelmed the early Indian English writers and so it did for the early Indian travel writers as well.

Swaralipi Nandi in her essay "When the Clown Laughs Back" further rallies the supremacy of the Western world by Indian writers and points to the travelogue of Nirad C. Chaudhury *Passage to England* (1960). She points to his overwhelming admiration for the English that overrides the narrative of his travelogue: "though he does critique the British…, the overriding narrative of his travelogue extols British culture and societal values as supreme and characterizing a higher civilization" (2014, p 5).

All these have become possible with the introduction of postcolonial studies, which has significantly opened the window for a contrapuntal reading of the early travelogues. In the essay titled "Tandoori Pizza: Fluid Culinary/Cultural Identities in Select Contemporary Indian Travelogues," Indrani M Desai analyzes the individual characteristics of postcolonial outlook:

> Postcolonial studies see early travelogues as narratives of power and desire. They see a strong historical connection of exploration with exploitation and occupation. Travelogues present not only the cultural identity of the land but also the cultural identity of the traveller. Postcolonial critics identified a monolithic Eurocentrism in travelogues. (2015, p 1499)

However, the above argument on early travel writings is not to dismiss travel narratives written around the same time that eyed the Western world through a critical lens. Works of writers such as that of Rabindranath Tagore *Letters from an Expatriate in Europe* (1888), Prafulla Mohanti's *Through Brown Eyes* (1989), and Bijoy Chand Mahatab's travelogue *Europey Tin Mash* (n.d.) are good examples. Krishnabhabini Dasi, in her book *Englande Banga Mahila* (Sen, 1996), portrayed the gloomy sides of England. Even though her encounter with the West started with an enthusiastic note, she portrayed the reality and recorded it in her travelogue which displayed her "distaste for the city's weather, the depression of nineteenth century industrialization and the gloom of the shabby shanty towns where poor workers crowded in squalor" (as cited in Nandi, 2014). Nabaneeta Dev Sen's *Dr Dev Sen's Foreign Trip* (Dev Sen, 1996) is another work that engages in returning the gaze. Nabaneeta Dev Sen in her works dismantles the long-established Eastern perspective while looking at the West and also skilfully uses the agency of laughter—an agency which was under the control of the Euro-American travelers for long. Such shifts in travel narratives are a welcome change that has allowed contemporary postcolonial travel writers to (1) recognize the space and narratives around them, (2) cognize the significance of possessing the narrative, and importantly, (3) write from an insiders' perspective.

8.3 SHIFT IN TRAVEL PERSPECTIVES

Bishwanath Ghosh, a Bengali travel writer and a journalist, joins his contemporaries Rajat Ubhaykar, Pankaj Mishra, and Monisha Rajesh as he charts a unique travel narrative that documents the experiences of journeying rather than writing elaborately about a specific destination, which has been the case for most travelogues. Monisha Rajesh's ambitious travelogue *Around the World in 80 Trains* (Smith, 2019) and Bishwanath Ghosh's *Chai, Chai: Travels in Places Where You Stop But Never Get Off* (Ghosh, 2009) offer similar narratives, as their works explore the journey that takes place inside the train as much as they take the readers through the less-explored landscapes. In a separate interaction, both these writers express a unique sense of community that forms in train journeys. The sense of romance associated with traveling by train, according to Rajesh, usually lives "in the passengers who would always tell their story to strangers, offer advice, share their food, and give up their seats" (as cited in Smith, 2019). In the prologue of his book, Ghosh mentions, "the journeys are not just about the levelling [the different class of people travelling in a close space], but also about getting acquainted with each other's cultures, especially food habits (2009, p 2)." An essay titled "Journeying through the Indian Railways" in *Around India in 80 Trains* (2012) by Monisha Rajesh and *Chai, Chai: Travels in Places Where You Stop but Never Get Off* (2009) by Bishwanath Ghosh interestingly equates the Indian train to the miniature version of a diversified India. The author Siddharth Dubey attempts to capture the specific details of people in the liminal, in between space (inside the Indian train) as documented in the travelogues:

(the Indian train reflects) a true sense of transient cultural pattern as it travels through different states of India constantly assimilating people of diverse cultures. In this liminal space, a passenger travels from known to unknown in terms of geography, culture, language, cuisine, sartorial configuration, and psychological makeup. Indian Railways offers an insightful analysis of cohabitation – the conflict and the coexistence of people amidst cultural differences. (2020, p 320)

Just as Monisha Rajesh encounters people of various cultures and backgrounds coming together in trains and the experiences of which she states, "trains are rolling libraries of information, and all it takes is to reach out to passengers to bind together their tales" (as cited in Smith, 2019), Bishwanath Ghosh takes over the train junctions/stopovers to understand the history, people, and the culture of small towns that remain overshadowed by bigger cities.

A similarity among these contemporary travel writers is the attempt to explore the unusual and less sought-after locations; however, intriguingly, locations that are important in terms of trade, commerce, and geographical connectivity with the rest of the country. This is why Ghosh's novel *Chai, Chai* serves as an important text in understanding a new perspective in travel narratives as he attempts to explore towns by rummaging important historical connections attached to these places. Ghosh's novel is an interesting exploration as he examines some of India's biggest railway junctions: Itarsi, Mughal Sarai, Jhansi, Saharanpur, Arakkonam, and Jolarpettai; places, which despite their significant roles in connecting towns and cities, are rarely in the itinerary of a traveler. Ghosh, through his work, attempts to re-define the idea and choices of travel destinations which are mostly associated with popular tourist locations and, hence, has allowed the country to be reductively defined through these selective places. Ghosh comments:

> ...these junctions, even though they bind the extreme corners of India, are hardly ever mentioned other than in the context of train travel. That is because as towns, they are too small to matter to you. They too must be having stories to tell- just that nobody ever steps out of the station yard to listen. (2009, p 4)

Ghosh begins his journey with Mughal Sarai where he attempts to confront the railway junction's long-term association as a recharging point for most travelers from the central to eastern part of India. Arindam Chakrabarti mentions this same association in his essay "Beranor

Dinbadal" (roughly translated as "Changing Days/Patterns in Travel"[2]) published in *Ananda Bazar Patrika,* a Bengali daily newspaper. He recalls Bengali travelers' expedition, in earlier times, on arriving at the Mughal Sarai junction early in the morning which signaled a meal that the family savored together during the trips. Arindam rekindles the refreshing image of Mughal Sarai which adds to Ghosh's recollections, "...the train would make an interminably long halt here during our annual trips from Kanpur to Calcutta. It was here that lunch would be served, in compartmented aluminium trays, along with tepid water... " (2009, p 8).

The novel records Ghosh's unconventional visit to capture the story of the towns behind some of the familiar railway junctions. In capturing the essence of the towns he visits, he also particularly comments on the significance of the railway junctions that often tends to fade in the familiar way it serves as just another "stopover." Ghosh says, "the railways are not just a means of transport, but the circulatory system of India. No Railways, no India. But not many spare a thought for the arterial valves that pump the blood: the big junctions which facilitate the movement of trains from one corner of India to the other" (2009, p 3). A noteworthy feature of Ghosh's work is in the way he complicates the idea of familiarity and re-produces it from a known, yet new, perspective; an immediate example are the railway junctions that are usual for travelers traveling from the eastern to the southern part of the country; however, junctions that have rarely sparked an interest among the commuters. There are several examples in the novel that confirm this technique deployed by Ghosh, one being the consistent mention of the crucial roles each of these junctions serves on the railway map. The selected sections from the novel helps us understand how Ghosh tackles this familiarity and its significance:

> It then struck me that nearly all trains running across the length and breadth of the country—Bombay to Calcutta, Delhi to Madras..., and so on—will have to pass through Itarsi...so crucial is its location on the railway map. Yet, Itarsi is almost non-existent in our daily life (2009, p 6). ...Guntakal, a railway junction on the south-western edge of Andhra Pradesh which, for decades, has been serving as the transit point between west India and south India (2009, p 137). ...Jolarpettai, or Jolarpet, the midway mark between Chennai and Bangalore and a key junction for all trains

---

[2]Roughly translated by the authors of the paper.

Journeying Through the Lesser-Known Indian Spaces
141

headed for Karnataka, Kerala, and southern Tamil Nadu …This is the convergence point of three important divisions of the railways … where the crew changes (2009, pp 177–186). … when South Indian Railway (a company that existed long before Independence) started a long-distance train from Madras to Mangalore [and later extended the service to Cochin], Shoranur happened to be in the middle of the route … That's when Shoranur became an important junction. (2009, p 197)

These important details are accompanied by the descriptions of the towns he visits. For instance, the simplicity of small eateries in towns where tables are shared and conversation naturally ensues between strangers, the shop names that carried the legacy of the family—"Krishnamurthy and Sons," "Lingamurthy and Co.," "Sivaram Traders" (2009, p 143), "Adarsh Vastu Bhandar" (2009, p 34), etc., and the endearing old-world-assurance that Ghosh observes between shopkeepers and customers. In doing so, he portrays an India which is "still very old India but much of India [is reflected in the simplicity that continues to exist in the towns despite the presence of globalization]" (Sen, 2021). In an interview with Amrit Sen, Ghosh expresses his idea about the real India; India that one gets to see when one travels 40–60 km from the brand name cities. This search for the "real India" surfaces in his writings and particularly in the novel *Chai, Chai* by using terms such as "Hindi heartland" (2009, p 16) and "Indian hospitality" (2009, p 24) to accentuate the essence of an Indianness that exists in the towns.

What Ghosh also does in writing about these places is re-defining the narrative of these locations that are often projected negatively due to criminal activities. His precise detailing and observation of the place and the people he encounters shows an effort he undertakes to understand the towns beyond its unpleasant stories. His own apprehension is visible as he steps out into the town of Mughal Sarai:

I had heard only unflattering things about Mughal Sarai. An editor, who had never been there, told me the place was crime-ridden. A police constable, who had lived there, warned me that the place was infested with *goondas*. (2009, p 7)

However, a close reading of the novel allows the readers to recognize his objective in the conversations he has with the people he meets and seeks to meet. This begins in the first town as he seeks for an answer to his query of "why was Mughal Sarai called Mughal Sarai" (2009, p 32). As he journeys further, he narrates the historical connections of the town—Lal Bahadur Shastri shares his birthplace with the town of Mughal Sarai, the renaming of the junction from Mughal Sarai to Pandit Deen Dayal Upadhyay as a tribute to the later. In Itarsi, he comes across a dharamshala that qualifies as an important piece of history, "Gandhi spent a night there—in 1933 ... Nehru had stayed here only for a few hours in 1937 ... and so had Rajendra Prasad, who had spent three hours here while on his way to Allahabad from Wardha" (2009, pp 115–117). Ghosh also records Shoranur town that stood on the banks of the Bharathapuzha, which he says is "Kerala's equivalent of the Ganga" (2009, p 191).

8.4 NEW TRAVEL PERSPECTIVES

Somdatta Mandal in her book *Indian Travel Narratives: New Perspectives* (Lahiri, 2021) presents a chronology of different kinds of travel narratives from India and explores the shift in the travelers' attitude from the 18th and 21st centuries. A section titled "Pilgrimages" from the book is an interesting inclusion, as it explores the travel history around the 18th and 19th centuries and also connotes an important, and often the sole, reason for travel during these periods for most Indians. In this context, Shobana Bhattacharji's observation in her essay "Indian Travel Writing" on religious traveling should be mentioned. She says, "Indian men and women have historically made journeys for reasons of religion [among other purposes]" (2015, p 125). She further comments on the growth and development of religious tourism in India as well. She records:

> In the first millennium BCE, moreover, Varanasi (Benares), was established as an important spiritual site for both Hinduism and Buddhism, attracting pilgrims from all over India and generating, as a consequence, several pilgrimage routes, and travel infrastructures to support them. Between the seventh and seventeenth centuries CE, the cults of Balajee in South India (Tirupati) and Jagannath (Puri) developed, and are still important pilgrimages for Hindus ... .(2015, pp 126–127)

The attitude toward traveling is taking a new turn in response to the COVID-19 global pandemic. In the literary circle too, an increasing

Journeying Through the Lesser-Known Indian Spaces 143

interest has been noticed in exploring travel narratives where writers put themselves in adventurous, challenging, and experimental roles. For Ghosh, it is the excitement in unplanned trips and traveling through unfamiliar places. Ghosh's attempt to write about these places is clear as he states: (his objective is) "to grab the reader gently by the arm and take them with me to the lands…while telling them the story ... This has been my approach for all my books" ("If Only People Travelled," 2018). Similarly, Rajat Ubhaykar's unusual travel expeditions in trucks across the country is note-worthy. In doing so, Ubhaykar contributes a new narrative in highway literature.[3]

Recognizing the future scope of travel writing at least for the Indian writers given the proportion and potential of the country, Talwar of Westland Books, a publishing house in India, says:

> As for me, my sincere hope is that in the years to come, we will find travelogues engaging with the local, books that explore the sub-regions, districts and tehsils of India's many states, each of which could qualify as a country by itself. I, for one, would love to know what the texture of life in Tikamgarh is like, what people eat for breakfast in Kendujhar, about the dejections and aspirations of the residents of Tamenglong. (The Future of Travel Books section, para. 5)

Rabindranath Tagore's lines "Dekha hoy nai chokkhu meliya/Ghar hote sudhu dui pa feliya … " (roughly translated as "But I haven't seen with these eyes/Just two steps from my home lies … ") (কবিতা ১ – একটি শিশির বিন্দু/Poem 1—Ekti Shishir Bindu (A Glistening Drop of Dew)", 2022) is a good reference which points to the unexplored spaces around us. The interest among travelers to explore local places is renewed more than ever before post pandemic with an intent among most people to look for nearest escape. The COVID-19 pandemic with its restrictions on movement and various other SOP's, leaves travelers with little choice than

---

[3]With writers like Rajat Ubhaykar, Indian travel writing is making its debut into the "highway literature"—a space essentially devoted to the lives and experiences on the roads; metaphorically a space devoted to the continuity of the traveling movement rather than the traveling destinations. With the observation that " … the critical difference between the truck drivers and bus drivers was that the bus driver goes home to sleep. The truck driver doesn't merely drive on the highway; the highways are where he lives out most of his life" (*Truck de India!* Rajat Ubhaykar, p. 10) and with an interest to document that living experiences of the truckers—travel writing in India is flourishing in a different dimension altogether.

144 Language and Cross-Cultural Communication in Travel and Tourism

to explore the local places. This is, however, not to dismiss the existing and renewed interests among travel enthusiasts to explore and understand the local geographies among. Aaron Gilbreath in his article "How Travel Writing May Look After the Pandemic" comments on the future of travel industry post pandemic:

> Many people have suggested that, once we are free from lockdown, more modest domestic or local travel, rather than exotic foreign adventures, will take center stage. They say narratives about home might become significant and popular. (2020, para 2)

Similarly, India's tourism sector has predicted to witness an increasing interest among Indian travelers to travel to the local places in the post-pandemic era. An article published by *The Economic Times* states a good possibility of people preferring to travel local in the immediate post COVID-19 phase, which would present a "huge opportunity for the Indian hospitality industry" ("Indians to Prefer Local Travel," 2021).

Latest studies on the changing modes of traveling after the global pandemic resonate with Gilbreath's observation. A study titled "The Shifting Trends in Travelling after the COVID-19 Pandemic," for example, reveals how travelers are becoming drifters[4] by nature. The survey result shows that:

> ... the respondents' answers were dominated by the drifter tourist typology, which was similar to the explorer type where they chose destinations they had never visited and travelled in small numbers. (Kusumaningrum and Wachyuni, 2020, p 35)

The same study reveals a shifting trend in travellers' attitude:

> On average, respondents agreed that after this Pandemic, they wanted to plan and create their itinerary and go on walks where they could interact with the environment and local communities.

---

[4]Cohen (1972) categorizes tourists into four different types:

(1) Organised mass tourist. (2) Independent mass tourist. (3) Explorer. (4) Drifter. The "Drifter" type of traveler, as defined by Cohen, is quite distinct from other travelers in that they seek for a solo experience, go to an unfamiliar location, make an effort to fit in, and behave as the locals would. They lack a set itinerary and definite destination. Although there is a high level of freshness, there is no amount of familiarity.

Furthermore, the respondents' answers were dominated by the drifter tourist typology, which was similar to the explorer type where they chose destinations they had never visited and travelled in small numbers. (Kusumaningrum and Wachyuni, 2020, p 35)

In a more geographically relevant essay "Post-Pandemic Travel: Decoding the Trends and Challenges for Indian Travellers," the study reveals the Indian travel patterns before and after the pandemic. The study observed at various parameters and interestingly, one of the variables showed an increase in domestic travel:

Domestic destination was the preferred choice of the majority of the respondents, especially for destinations such as Ladakh, Goa, North East region, Kerala, Himachal Pradesh, Kashmir, and Bhutan being not so far behind in the Indian sub-continent. Self-driven destinations with reasonable distance were also preferred, such as Coorg, Ooty, Mussoorie, Shimla, Amritsar, Munnar, etc. (Sneha, 2021, p 81)

However, the COVID-19 pandemic with its restrictions and the fear surrounding it raises concern for writers such as Ghosh and Ubhaykar as their approaches engage with first-hand interactions with people. Rajat Ubhaykar comments on this:

Travel for me is about people and not places, and I am worried that the unmitigated pleasure of conversing with complete strangers and learning about each other's disparate lives is going to be marred by suspicion and paranoia. (The Future of Travel Books section, para 2)

8.5 CONCLUSION

The introduction to Indian railways, the growth in science and technology, and the advent of modernity contributes to the shift in travel attitudes in the recent past; from religious travel to travel for leisure or pleasure travel. The current phase in tourism has been initiated by the young Indian travelers, for whom traveling is more about a secluded solo experience and pure adventure; as opposed to the common interest of earlier generations of travelers invested in traveling to religious places. The young Indian travelers prefer roaming around the beaches, trekking on hills, rafting on rivers, and paragliding, etc. Most importantly, they show a keen interest in exploring the offbeat Indian places (for example, the underhyped beaches like Bakkhali beach, Taal sari Beach, and Rishi Konda beach, etc.). This shift has influenced contemporary travel writings and has rightly been observed by Somdatta Mandal in her book *Indian Travel Narratives: New Perspectives* (Lahiri, 2021). Himadri Lahiri in a review for Mandal's book states, "While the earlier travel writers recounted the experience of their pilgrimages, the later writers usually responded to the call of the wild and sublime beauty" (2021, p 209).

Travel writing from India has taken interesting turns, from its earliest writings which were mostly documentations of travel for administrative and commercial reasons, to the recent ones that are more exploratory and adventurous in nature. From writing to reverse the colonial gaze to the recognition of the significance in writing from an insider's perspective, Indian travel writers have marked their nascent yet powerful presence in the genre of travel writing. This marks a slow yet gradual departure from the colonial hangover reflected in the early English travel writings.

KEYWORDS

- **Indian travel literature**
- **geohumanities**
- **space**
- **local travel**
- **representation**
- **train travelogue**
- **diverse culture**
- **pandemic**
- **travel**

REFERENCES

Banerjee, R. In *India in Early Modern English Travel Writings: Protestantism, Enlightenment, and Toleration*; Brill, 2021. https://doi.org/10.1163/9789004448261

Bhattacharjee, S. Indian Travel Writing. In *The Routledge Companion to Travel Writing*; Thompson, C., Ed.; Routledge, 2015; pp 125–138.

Chakrabarti, A. বেড়ানোর দিনবদল. *AnandabazarPatrika* [Online] 2021. https://www.anandabazar.com/rabibashoriyo/history-of-tours-by-bengalis/cid/1306883

Chatterjee, A. C. Maharaja Adhiraja Bijay Chand Mahtab of Burdwan. *J. Asiatic Soc.* **2011,** *73* (2), 387–388. https://doi.org/10.1017/S0035869X00097835

Desai, M. I. Tandoori Pizza: Culinary/Cultural Identities in Select Contemporary Indian Travelogues. *Eur. Acad. Res.* **2015,** *3* (2), 1491–1512. https://www.euacademic.org/UploadArticle/1624.pdf

Dev Sen, N. Dr Devsener Bidesh Yatra. [trans. Dr Dev Sen's Foreign Trip]. In *Nabaneeta Devsener Galpo Samagra: Part 1*; Dey's Publishing, 1996; pp 28–45.

Dubey, S. Journeying Through the Indian Railways in Around India in 80 Trains (2012) by Monisha Rajesh and Chai, Chai: Travels in Places Where You Stop but Get Never Off (2009) by Bishwanath Ghosh. *Rupkatha J. Interdiscip. Stud. Humanit.* **2020,** *12* (3). https://doi.org/10.21659/rupkatha.v12n3.38

Ghosh, B. In *Chai, Chai: Travels in Places you Stop but Never Get Off*; Tranquebar Press, 2009.

Gilbreath, A. How Travel Writing May Look After the Pandemic; *Longreads*, 2020.https://longreads.com/2020/06/02/how-travel-writing-may-look-after-the-pandemic/

If Only People Travelled, India Would be more Peaceful; *Manorama*, 2018. https://www.onmanorama.com/travel/outside-kerala/2018/08/19/if-people-travelled india-would-be-more-peaceful.html

Islam, M. Md.; Das, S. Travel Writing and Empire: A Reading of William Hodges's Travels in India. *PostScriptum: Interdiscip. J. Lit. Stud.* **2017,** *2* (3), 1–15. ISSN: 2456-7507

Kusumaningrum, D. A.; Wachyuni, S. S. The Shifting Trends In Travelling After The Covid-19 Pandemic. *Int. J. Tour. Hosp. Rev.* **2020,** *7* (2), 31–40. https://doi.org/10.18510/ijthr.2020.724

Lahiri, H. Review of Somdatta Mandal, ED. Indian Travel Narratives: New Perspectives. [Review of the book *Indian Travel Narratives: New Perspectives*, by Somdatta Mandal]. *Indi@logs* **2021,** *8*, 209–214. https://doi.org/10.5565/rev/indialogs.187

Mandal, S. Ed. In *Indian Travel Narratives: New Perspectives*; Pencraft Publication, 2021.

Mohanti, P. In *Through Brown Eyes*; Oxford University Press, 1985.

Mukherjee, T. N. In *A Visit to Europe*; Newman, W., 1889.

Mukherjee, M. In *The Perishable Empire: Essays on Indian Writing in English*; Oxford University Press, 2000. 10.2307/23341547

Naidu, V. Indians to Prefer Local Travel in Post-Pandemic Phase; *The Economic Times*, 2021. https://economictimes.indiatimes.com/news/politics-and-nation/indians-to-prefer-local travel-in-post-pandemic phasevenkaiah naidu/articleshow/80231404.cms?utm_source=contentofinterest&utm_medium=text&ut m_campaign=cppst

Nandi, S. When the Clown Laughs Back: Nabaneeta Dev Sen's Global Travel and the Dynamics of Humour. In *Studies In Travel Writing*, 2014; vol *18* (3). https://doi.org/10.1080/13645145.2014.942102

Nayar, P. Marvelous Excesses: English Travel Writing and India, 1608–1727. *J. Br. Stud.* **2005,** *44* (2), 213–238. https://doi.org/10.1086/427123

Nayar, K. P. In *The Colonial Voices: The Discourse of Empire*; John Wiley & Sons, 2012.

Rajesh, M. In *Around India in 80 Trains*; Bloomsbury Publishing, 2019.

Said, E. In *Orientalism*; Penguin India: London, 2001.

Santhanam, R. Ethnocentrism. *The Hindu*, 2022. https://www.thehindu.com/society/ethnocentrism/article38369090.ece.

Satapathy, A. Reconsidering the West in Early Autobiographies and Travel Writings in Indian Writing in English. *J. Law Soc. Sci. Singapore* **2012,** *2* (1), 177–182. ISSN 2251- 2853.

Sen, S. Ed In *Englande Banga Mahila*; Stree, 1996.

Sen, A. Chai, Chai: Interaction with Bishwanath Ghosh [Video]. YouTube, 2021. https://youtu.be/Sak-qFSyg0U

Shiels, S. What's it Like to Travel the Entirety of India (or 26 of 29 States)? [Online] 2016. Quora.com.https://www.quora.com/Whats-it-like-to-travel-the-entirety-of-India-or-26-of 29-states

Smith, P. D. Around the World in 80 Trains by Monisha Rajesh Review- The Romance of Rail Travel; *The Guardian*, 2019. https://www.theguardian.com/books/2019/jan/26/around-the-world-in-80-trains-by-monisha rajesh-review

Sneha, N. Post-Pandemic Travel: Decoding the Trends and Challenges for Indian Travellers. *Atna J. Tour. Stud.* **2021,** *16* (1). https://doi.org./10.12727/ajts.25.4

Tagore, R. In *Europe-Prabasir Patra*; Tagore, R., Ed.; Digital Library of India, Free Download, Borrow, and Streaming: Internet Archive, 2005.

Theroux, P. In *The Great Railway Bazar*; Houghton Miffin Harcourt, 1975.

Ubhaykar, R. In *Truck de India! A Hitchhiker's Guide to Hindustan*; S&S India, 2019.

Ubhaykar, R. Where is Indian Travel Writing going after the pandemic (and where is it coming from)? Scroll.in., 2020. https://scroll.in/article/972052/where-is-indian-travel writing-going-after-the-pandemic-and-where-is-it-coming-from

কবিতা ১ – *একটি শিশির বিন্দু* / Poem 1 – Ekti Shishir Bindu (A Glistening Drop of Dew), 2022. https://jyotirjagat.wordpress.com/2014/06/15

CHAPTER 9

Imagined Communities: The Development of the Early Tourism Industry in Alaska and the Marketing of the Indigenous Experience

VERA PARHAM and JENNIFER WILLIAMS

Department of History, American Public University, Charles Town, WV, USA

ABSTRACT

As the tourist and cruise industries developed, Americans with new middle-class wealth hoped to visit distant lands and people and experience exotic cultures. The territory of Alaska proved to be a popular destination for providing this experience. White middle-class Americans could explore the wild wonders of sun and snow and bring a small piece of their adventure home by purchasing Indigenous symbols and souvenirs. The production and commodification of these souvenirs, for the entertainment and education of tourists, had a direct impact on not only the white concepts of indigeneity and "nativeness" but also on the preservation and perceived authenticity of indigenous cultures. The commodification of Alaskan Native Arts represents the ability to preserve material culture for Alaskan Natives and can be interpreted as both tools for self-representation and involvement in the economy, as well as a whitewashing of Indigenous lifeways. This paper focuses on the growth of the cruise industry in southeast Alaska from 1870 to 1940 and the development of the curio trade to

---

Language and Cross-Cultural Communication in Travel and Tourism: Strategic Adaptations.
Soumya Sankar Ghosh, Debanjali Roy, Tanmoy Putatunda, & Nilanjan Ray (Eds.)
© 2025 Apple Academic Press, Inc. Co-published with CRC Press (Taylor & Francis)

150 Language and Cross-Cultural Communication in Travel and Tourism

increase tourist traffic to the region through the lens of curios shops, cruise line promotions and published journals of travelers to Alaska.

9.1 INTRODUCTION

Tourism and travel. The words conjure up a sense of adventure and excitement. The romance of discovery and uncovering new lands and people. They represent the ability to find ones way out of the mundane of daily life and routines and to experience the world from a new perspective. They also represent the opportunity to interact with and experience new cultures. In order to transport tourists out of their daily lives, the industry relies upon marketing and selling an experience. And, in order to do that, the experience must be carefully packaged and presented—a curated vision of a supposed authentic experience. In tailoring and creating this touristic experience, the industry inadvertently relies upon cultural exploitation yet also promotes transcultural communication through fostering the roles of the consumer and the consumed. The experience can inform the consuming participant about the viewed culture, but often says more about the ideas and desires of those doing the consuming. This study is a mixed-methods cultural exploration related to the historical interpretation of the material culture of the tourism industry in late-19th century Alaska and will focus on the development of the cruise industry in that area and the commodification of Indigenous culture for mass appropriation and consumption. One of the key research questions is to uncover how and if the propagation of Indigenous crafts and the creation of "imagined communities" through the marketing of the exotic and the other represented an opportunity to disseminate authenticity. The study examines the "white gaze"—how the tourist industry, collecting and cruise trips informed the white perspective of "Nativeness"[1] with an eye toward Alaskan Natives as active participants and stakeholders in an industry often controlled by outside corporations.

The region, now known as Alaska, has long Native traditions that predate Western contact.[2] These traditions, practices, and beliefs varied depending on location. Divided up by their language, the earliest European

---

[1]See: Henay, C. There is Nowhere to Hide: Spirit and Heart in Afro-Indigenous Transformative Engagement. *Cultural and Pedagogical Inquiry,* University of Alberta, Fall 2018, *10* (2), 1–4.

[2]See: Haycox, S. W.; Mangusso, M. C. In *An Alaska Anthology: Interpreting the Past*; University of Washington Press: Seattle, WA, 2011.

Imagined Communities: The Development of the Early Tourism Industry 151

contact was with the Aleuts of the chain of islands that bear their name, as well as with the Tsimshian and Haida of northern British Columbia and southeast Alaska. As the last Indigenous group in North America to sustain prolonged if sporadic contact with colonial empires, their culture remained relatively intact, although the fur trade and settlement would begin to change traditional practices. Before the ferries and the barges and fishing vessels and cruise ships, and even the European explorers plied these waters in the 17th and 18th centuries, Alaskan Natives navigated the region in canoes, both for trade and war. In a prophetic view of the significance sea vessels would have on the alteration of Alaskan Native life, Chief Toy-a-Att, a Sitka Indian said, "White men appeared before us on the surface of the great waters in large ships which we called canoes. Where they came from, we knew not, but supposed that they dropped from the clouds."[3] The Tlingit, Haida, and Tsimshian developed distinctive cultures with a society based on reciprocity and a reliance on their environment for survival. Salmon was dried and smoked and wild berries were gathered and dried and made into cakes for winter; bears, deer, and beaver were hunted for hides and meat. They utilized cedar and spruce trees for making baskets, regalia, canoes, and one of the most identifiable symbols of Southeast Alaskan culture, the totem pole.

On the other hand, the leisure travel industry is but a century and a half old.[4] Taken from the 18th-century concept of the Grand European Tour of the very wealthy and the aristocracy, which has its antecedents in the pilgrimages and exploration of the Medieval and Renaissance periods, leisure travel became a marketable industry in the United States after the Industrial Revolution in the 19th century.[5] As more people immigrated or moved into the cities, the populations swelled into factories and industry. The new work schedule, consisting of five days a week and two days for leisure time, allowed for the rising middle class to take time off for travel. This new urbanization and modernization influenced the longing for and romanticizing of sites such as Alaska, still wild and free. Railroads hauled

---

[3]Chief Toy-a-Att in Meaux, J. M. In *Pursuit of Alaska: An Anthology of Travelers Tales*; University of Washington: Seattle, 2013; p i.

[4]Campbell, 6. Campbell Briefly Discusses the Development of the Tourism Industry and the Lens of 19th and Early 20th Century Traveler to Alaska in his First Chapter in "In Darkest Alaska", which is an Allusion to the Rudyard Kipling Poem. Campbell, B. In *Darkest Alaska: Travel and Empire Along the Inside Passage*; UofPENN Press: Philadelphia, 2007.

[5]Walton, J. K. In *Histories of Tourism: Representation, Identity and Conflict*; Channel View Publications: Clevedon, 2005; pp. 4–8.

152 Language and Cross-Cultural Communication in Travel and Tourism

freight and paying passengers into parts of continents still little settled by Euro-Americans or rarely (if ever) seen by outsiders.[6] Eventually, steamship travel and trans-oceanic luxury liners crossed the open waters, carrying passengers and freight to destinations across the globe. By the middle of the 19th century, established travel tours were available. The collection and display of travel souvenirs became a fashion statement in middle class Victorian era homes as a symbol of the exchange of goods related to the travel industry.[7] Steamship companies advertised trips to exotic-sounding places, cottage hotels sprung up along the roadside for drivers weary of the road and tourist wares, a souvenir of your experiences and place of travel, were commonplace.

As the tourist and cruise industries developed, Americans with new middle-class wealth hoped to visit distant lands and people and experience exotic cultures. The territory of Alaska proved to be a popular destination in providing this experience. It took up the mantle of the "Old West" and came to represent the fading gasp of Manifest Destiny. The concepts of space and place are incredibly significant when looking at the tourist industry.[8] The industry markets space and transports individuals to a specific place. But those places are also the foundations of Indigenous identity and so to market the place, it becomes convenient to market the indigeneity of the place. Cultural items occupy a specific place in their respective communities and their meaning or significance shifts when it is coopted or marketed for touristic purposes. Thanks to the marketing strategies of combining the wild (nature) with the exotic (Indigenous) into a comfortable tour on board a modern cruise ship, white middle class Americans could explore the wild wonders of sun and snow and bring a small piece of their adventure home through purchasing Indigenous symbols and souvenirs. As well as the menus, postcards and other ephemera embed with Indigenous images. The production and commodification of these souvenirs, as well as the later performance of cultural practices for the entertainment and education of tourists, had a direct impact on not only the white concepts of indigeneity

---

[6]The Development of Transportation is in Direct Correlation with the Development of Tourism as one can't be a Tourist Unless One Travels! See: Likorish, L. J.; Jenkins, C. L. In *An Introduction to Tourism*; Routledge: London, 1997; pp. 11–17.

[7]Campbell, 6.

[8]Bunten, A. C. More Like Ourselves: Indigenous Capitalism Through Tourism. In: *American Indian Quarterly,* Summer 2010; vol *34* (3), p. 285.

Imagined Communities: The Development of the Early Tourism Industry 153

and "nativeness" but also on the preservation and perceived authenticity of indigenous cultures themselves.[9]

While all tourism focuses on marketing and imagery, in Alaska, that imagery was focused almost entirely on the wild nature and Indigenous roots of the region. Instead of a priceless Greek sculpture or Venetian glass, the tourist trinkets were more affordable, accessible, and more easily produced for mass consumption to match the growing numbers of tourists. This process required the commodification of historic sites to make the experience more tangible.[10] The changes in transportation technology and the rise in leisure time and middle-class incomes in particular in the late-19th century contributed to the rise of the leisure travel industry for more than just the wealthy, it contributed to the changes colonization wrought on Indigenous cultures. Colonial contact is generally destructive for Indigenous cultures, especially when it targets the exploitation of natural resources, such as Russian fur harvesting. However, many social scientists feel that tourism actually encourages the preservation and revitalization of culture through the marketing of arts and crafts and the educational use of historical sites.[11]

The production of curios, items for the tourist trade, has a long history on the Northwest Coast. In 1791, Alejandro Malaspina visited Yakutat Bay and wrote that the locals, "discovered the market for (figurines, spoons, daggers, boxes etc.) ... and began to make them for trade."[12] Alaska had a small trade in native goods produced by the Native groups themselves, such as the Tlingit co-ops. These co-ops were generally informal groups of women looking for ways to practice traditional arts while supporting their families financially. For many Southeastern Alaskan Native women,

---

[9]The dances would not be organized into events until after World War II for the most part. Some dances were performed for tourists but they were not a regular event. Items for tourists and the imagery of "nativeness" associated with Alaska are separate from the "made for sale" type of art that was systematically collected by those like George Gustav Hye and addressed in the seminal work of Nelson Graburn In *Ethnic and Tourist Arts: Cultural Expressions from the 4th World*; University of CA Press: Berkeley, 1976.

[10]Hill, L. L. Indigenous Culture: Both Malleable and Valuable. *J. Cult. Herit. Manag. Sustain. Dev.* 2011, *1* (2), 122–134.

[11]See; Boissevain, J. In *Coping with Tourists: European Reactions to Mass Tourism*; Berghahn Books: (NY, 1996. Emanual DeKadt, *Tourism: Passport to Development?* Oxford University Press: Oxford, 1979. Mansperger, M. C. Tourism and Cultural Change in Small-Scale Societies. *Hum. Organization* *54* (1), 87–94.

[12]From, de Laguna, F. Under Mt. St. Elias: The History and Culture of the Yakutat Tlingit. In *Smithsonian Contributions to Anthropology*, 1972; vol 1, p. 144.

154 Language and Cross-Cultural Communication in Travel and Tourism

participation in the pre-colonial and even colonial economy was encouraged and considered necessary.[13] The production of these goods, while representing traditional arts, were generally made out of nontraditional materials and thus allowed the artists a freedom to produce and promote their art in a new way.[14] The early curio trade, controlled by the Tlingit and Haida, responded to the uptick in visitation especially after 1880 producing a changing array of tourist items including beadwork and clothing.[15] Baskets were coveted possessions for Victorian travelers into the early 20th century, but small totem poles, moccasins, and carved ivory were also in demand.[16] By 1890, a report on Alaska's tourist trade stated, "the people of several villages devote themselves exclusively to the manufacture of curios ... "[17] Items were generally produced by willing individuals, but the painful legacy of the missionary-industrial schools also left a mark on the curio trade. In 1878, Sheldon Jackson helped find one of the first Protestant missionary schools in Sitka. In an attempt to foster financial self-sufficiency school attendees created curios for tourists.[18] For the western educators, these items served a dual purpose of bringing students into the colonial capitalist system while developing a marketing brand and imagery to promote the mission of the schools.

Alaska, like many tourist locales, experienced the commodification of representative icons (totems, masks, baskets, clothing, etc.). But it is not just the items for sale that encouraged tourists to visit these exotic lands. John Muir spurred an influx of tourists after 1879 with his vibrant descriptions of Alaskan life. There was also the imagery, and when the Kodak camera hit the market at an affordable price, the use of portable cameras to capture the romantic aspect of travel and the "locals" in sites in Alaska went hand-in-hand with the development of the tourist trade and the cruise

---

[13]See; Shortridge, F. Life of a Chilkat Indian Girl. *Mus. J. 4* (3), 101–103

[14]Lenz, M. J. Material: George Heye and his Golden Rule. In *American Indian Art Magazine*; Autumn 2004.

[15]See; Smetzer, M. In *Painful Beauty: Tlingit Women, Beadwork and the Art of Resilience*; University of Washington Press: Seattle, 2021.

[16]Campbell, 162–164.

[17]Porter, R. D. In *Report on Population and Resources of Alaska and the Eleventh Census: 1890*; Government Printing Office: Washington, DC, 1893.

[18]See; Carlton, R. In *Sheldon Jackson: The Collector*; Alaska State Museums: Juneau, 1999.

Imagined Communities: The Development of the Early Tourism Industry   155

industry.[19] And finally, there was the knowledge that Alaskan Natives were hospitable and friendly to outsiders.[20]

Population statistics for Alaska illustrates the financial boon and strain of a tourist economy. The 1880 census lists only 435 non-Natives in the region.[21] In 1890, after a devastating pandemic, the population of Alaska Natives of 23,531 was still larger than non-Natives at 4298.[22] By 1910, the population of non-Alaska Natives has increased by a hundredfold.[23] While there are no good statistics on the number of travelers each year to Southeast Alaska from 1870 to 1940, some details emerge from passenger lists and various accounts. It is safe to assume that it went from a handful each year in the first decade to several thousand each year by 1940. Over 5000 visitors came to Alaska in 1890.[24] Enthusiastic newspaper articles would tout the number of visitors to port, especially in Skagway and Juneau. Some statistics exist from the 1930s. From *Facts About Alaska*, the total passengers in 1933 was 16,117 and in 1934, it was 24,009. There is no distinction between visitors (often called round trippers in cruise ship documents) and residents.[25] The efforts of the Territorial Chamber of Commerce to develop a comprehensive and sustained marketing strategy was "hampered by lack of funds" so the "burden of advertising the territory falls upon the various transportation companies operation in Alaska."[26] The Wrangell Chamber of Commerce published a four-page brochure highlighting the commercial aspects of Wrangell but also dedicated space to present excursion options for visitors. *Stroller's Weekly*, a publication from Douglas Island, reported from the Alaska Bureau of the Seattle Chamber of Commerce that "the Pacific Steamship Co. and Alaska Steamship Co. are running all boats assigned to the Alaska run. Every ship leaving this port (Seattle) for the

---

[19]Meaux, J. M. In *Pursuit of Alaska: An Anthology of Travelers' Tales, 1879-1909*; University of Washington Press: Seattle, 2013; p xiv.

[20]Morgan Meaux, xv.

[21]Morgan Meaux, xiii.

[22]Morgan Meaux, 276.

[23]Morgan Meaux, xiv.

[24]Campbell, 6.

[25]*Facts About Alaska* 1935, 46.

[26]*Facts About Alaska*, 1935; vol 49.

156   Language and Cross-Cultural Communication in Travel and Tourism

Territory is booked to capacity."[27] The Alaska Steamship Company had 16 ships, seven of them operating in Alaskan waters by 1940.[28]

9.2   THE CRUISE INDUSTRY

The growing press and attention to Alaska in the late 1800s prompted the development of the cruise industry. The first registered cruise to Alaska was planned by railroad tycoon Henry Villard in 1881 and consisted of 80 passengers sailing through the Inside Passage.[29] The Pacific Steamship Company shifted from commercial to pleasure ships and advertised in *The Alaskan*, Skagway's daily paper, for the Steamer *Idaho*'s voyage from Portland, Oregon to Port Townsend, Washington Territory, Victoria and Nanaimo, British Columbia, and the ports of Wrangel (sic), Juneau, Killisnoo, and Sitka, Alaska.[30] There were five cruise ship lines that served Southeast Alaska by the 1920s although Alaska Steamship Company and the Canadian Pacific Railroad Princess ships brought in the most passengers.[31] The others were Humboldt Lines, The Pacific Steamship (also known as the Admiral Line), and the Grand Trunk Line. The cruise ships not only took passengers but freight, and not all passengers were "round trippers." In general, cruise ships itineraries into the 1900s included Metlakatla (near present-day Ketchikan), Wrangell, Sitka, the gold quartz mines at Juneau and Douglas Island, and Skagway. The trips might stop in the Glacier Bay and Ketchikan areas to visit canneries in the first decade of the 1900s. By the 1910s, Petersburg would be added to the itineraries and Metlakatla would be dropped. As the Victorian emphasis on collecting baskets diminished by the end of the 19th century, and more excursions were added to promote tourism to the region, towns would adapt by exploiting the natural wonders (Denver Glacier in Skagway and the Stikine River in Wrangell). Juneau would add trips to Auke Lake and Mendenhall Glacier as excursions by the 1920s. The Princess Lines included Alert Bay, British Columbia.[32] Skagway already had the White

---

[27]*Stroller's Weekly,* July 16, 1921.

[28]Alaska schedules, 1941, Alaska Steamship Company.

[29]Morgan Meaux, 5.

[30]*The Daily Alaska*, Dec 5, 1885.

[31]*Steamships*. The Daily Alaskan; Jan 1, 1918; vol 3.

[32]*Princess Steamships to Alaska*. Schedule; Canadian Pacific **Brochure, Alaska, 1929.**

Pass & Yukon Railroad, which offered whistle stops to Denver Glacier and daily trips to the White Pass summit. In addition, Skagway was able to offer additional excursions to Atlin Lake via the train in the 1920s and 1930s.[33] Haines was added by the Alaska Steamship Company by the 1930s.[34] Sitka would diminish in importance once the state capital was relocated to Juneau. However, Sitka would continue as a major excursion stop for the cruise ships.

A major draw for cruise visitors were totem pole parks. There were several places where totem poles were highly visible including Ketchikan, Wrangell, and Sitka so they were frequently visited by the cruise ships. The totem pole parks represent the dichotomy of tourism as they were initially created out of poles left standing in villages decimated by population loss (disease and migration) from colonization, but were also places where traditional pole carving was revived and thrived carrying forward Indigenous family stories. They represent the continued claims to land and heritage of the Tlingit and Haida, parsed through the gaze of the tourist onlooker. With funding from the New Deal programs of the 1930s, the parks were expanded to make "the outstanding evidences of craftsmanship of its native people"[35] available to a wider tourist audience. Today these parks are still a major draw and part of the route of most cruise lines— Ketchikan, Juneau, Skagway, and sometimes Sitka and Glacier Bay.

Much of the promotion for tourism came from the cities on the cruise routes and by the end of the 19th century, American perceptions of Alaska began to shift from an unknown wilderness to an exciting frontier due in large part to local promotion, newspaper coverage, and published travel journals.[36] By the 1920s, the individual towns began to write and distribute booklets to encourage economic development rather than tourism, but they did extoll the natural wonders of each location. The Skagway Citizen's Committee booklet, published in 1924, said "that Skagway is the best town for the Tourist to stop over in and Spend a Few Days."[37] Also discussed were the modern amenities of "electric lights—and good

---

[33]*Alaska, Atlin, and the Yukon.* In *Cruising To Alaska*; Alaska Steamship Company, 1935; Alaska Steamship Company, 1926; pp. 1–12.

[34]*Sailing News.* Alaska Steamship Company, July 2, 1939; n.p.

[35]Heintzleman, B. F. Indian Antiques S.E. Alaska to be Preserved by Forest Service; Wrangell Sentinel, 11/4/38; p. 1.

[36]Morgan Meaux, xiv.

[37]Skagway Citizen's Committee, 1924, n.p.

158 Language and Cross-Cultural Communication in Travel and Tourism

automobile service" since there was a Ford dealership in town. "There are also a number of Curio shops in Skagway, which are the largest and most up to date in Alaska…a source of great interest to the visitor."[38]

The description of Southeast Alaska travel brochures was designed to appeal to the "primitive" details and scenery of Alaska. Travel journals and travel accounts, from authors and seasoned travelers such as Septima Collis and Eliza Ruhman Scidmore, were a more personal look at cruising in Southeast Alaska. The journals vary widely in description and details and sometimes reveal the natural prejudices of the writer. Most journals were kept by women although not all. Edward Parkinson published his account in 1892 called—*Twelve Weeks—Wonderland* as he traveled on the *Queen* of the Pacific Steamship Company. Some were published and others were turned into fictional accounts of Southeast Alaska.

9.3 THE WHITE GAZE

The travel journals examined have many similarities across time. Most of the journals examined are from the 1920s and 1930s but there are some accounts available from earlier visits. Descriptions, both light and detailed, tend to discuss the scenery, and the weather foremost. Additional descriptions include the ports visited and their environs, places visited while ashore, and encounters with Native groups. The discussion of locations would change as cruise ships added a more detailed itinerary throughout the early 20th century. Where earlier stops in Southeast Alaska during the last quarter of the 19th century might have included Wrangell, Juneau, and Sitka, as the 20th century dawned, Skagway, Petersburg, and Ketchikan would be added as their importance as settlements grew. In addition, trips to the Yukon Territory and upper British Columbia were developed after the Klondike Gold Rush of 1897. By the 1920s, tours to Mendenhall Glacier were advertised after the construction of Glacier Road was completed. So, the variety of shore excursions expanded as enterprising individuals and businesses took advantage of the increasing numbers of travelers to Southeast Alaska through the time period. The discussions of Native groups in Southeast Alaska vary depending on the author. Some are derogatory, as in Septima Colllis's and Edward Parkinson's details of Wrangell and Sitka Tlingit, yet others describe their work ethic such in

---

[38]Skagway Citizen's Committee, 1924.

Imagined Communities: The Development of the Early Tourism Industry   159

relation to a tour of Sheldon Jackson's workshop and school in Sitka.[39] Yet, this discussion is in relation to Western standards of work and hygiene. In short, by our standards, some of the descriptions of the Tlingit and Haida are disparaging and insulting. Many interactions were cursory at best—women selling trinkets on the dock, a tour of a school, visiting a Tlingit home. These journals though are good accounts of the active curio trade that existed in Southeast Alaska and even mention some of the more well-known businesses and their owners, such as Princess Tom of Sitka.

The travelers here, Lilian Manberg Bundy, Mary Otis Miller Hayes, Septima Collis, Edward Parkinson, and Florence Carr Wendt, visited Southeast Alaska at different times and had differing experiences, leaving behind some account of their travels. Collis and Wendt, separated by almost 50 years in their travels, provided the most detailed accounts of their trip and both women published their travel journals. These few accounts barely scratch the surface of the depth and breadth of travel information to be gleaned from travel journals. But they are hard to locate and often scanty in some details. Mary Hayes's account is cursory at best. Yet, even these brief descriptions are valuable to understand the changes in the tourism industry between 1870 and 1940 and the importance of Native Artisans and the curio trade.

Mary Otis Miller Hayes, wife of Colonel Webb Cook Hayes (a son of Rutherford B. Hayes), took a cruise in July 1916 on the *S.S. Alameda* with her husband and two teen-aged nephews. She described Ketchikan, their first stop, as it "was raining true to the reputation that it always rains in Ketchikan."[40] During their two-hour visit, they "walked through the streets ... and visited an interesting curio shop where I bought a pen holder and a paper cutter decorated with designs of some historic totem poles."[41] As the ship continued northward, Wendt describes the scenery as "mountains ... straight up for three of 4000 feet and more—backs that are shaggy with Sitka spruce, and red and yellow cedar."[42]

The next stop on the cruise route was Wrangell. Bundy stated that "we went to see Chief Shakes house."[43] Chief Shakes personal home

---

[39]Collis, 82–83; Parkinson, 179–182; Campbell, 148–150.

[40]Hayes, p. 200.

[41]IBID

[42]Wendt, p. 15.

[43]Bundy, n.p.

160 Language and Cross-Cultural Communication in Travel and Tourism

was a frequent stop for travelers to Wrangell because of his collection of ceremonial regalia. After this brief visit, Bundy and her traveling companions "took in a curio shop ... The Indians displayed baskets and moccasins."[44] There was only one curio shop in Wrangell, the Bear Mountain Shop. Wendt referred to Wrangell as "the city of (to)tem poles."[45] She commented on the honesty of one of the shopkeepers who pointed out the difference between locally made wares and those that were made in Japan. There are several potential reasons why imported goods came from Japan starting in the 1910s and teens. A major flu epidemic decimated the Native population, making the production of goods for the tourist trade difficult.[46] Or simply the demand outstripped the supply and shopkeepers began to import goods made from elsewhere. It was still flourishing by the time that Florence Wendt traveled in the mid-1930s on the S.S. Yukon. It was a large enough trade to impact Native goods as evidenced by an attempt in 1921 to protect the Native Art trade. In an editorial supporting pending state legislation to protect Native craftsmen in Skagway's *The Daily Alaskan* dated January 11, 1921, it discusses the plan of protection for the Native Curio Art. The article states that "miniature totem poles are being manufactured in Seattle and sent to Juneau" to "palm them off as native workmanship. A great deal of ivory work is the product of Japanese handicraft."[47] Legislation was introduced to protect both the consumer and the Indigenous craftsmen and was welcomed "as a protection for the industrious natives but to protect the native art of this Territory ... the marking of each article so as to distinguish it."[48]

Septima Collis traveled to Alaska in the late 1880s. Along with Eliza Scidmore, the women produced two published works on Alaska that were considered early tour books of the Southeastern region. While the ideology and prejudices are intact in their works, they are nonetheless important to the understanding of travel in the region. Collis's travel journal, which was published in 1890, contains descriptions and wood block carving prints of locations throughout Southeast Alaska but especially Wrangell and Sitka. She describes the "fourth day out from Tacoma (June 5th) we

---

[44]IBID

[45]Wendt, p. 16.

[46]It is estimated that a flu and measles epidemic killed up to a quarter to a third of the Indigenous population in Alaska from 1900 on. In Meaux, p. 276.

[47]*The Daily Alaskan*, p. 3.

[48]*The Daily Alaskan*, p. 3.

Imagined Communities: The Development of the Early Tourism Industry 161

found ourselves at…Fort Wrangell"[49] and her thoughts on the small settlement of miners and military states that "Fort Wrangell is perhaps today as uninviting a sport as nay in the worlds, save for the few curiosities in the way of Indian graves and totem poles, and the excellent work being done by the missionaries at the Indian schools."[50] Wrangell was not touted as a place for visitors yet its location near the mouth of the Stikine River and its importance to the Cassier gold rush meant that a visit was in order. Collis does disembark, "picking my way as best I could through the muddy thoroughfare to get a view of my first totem pole."[51] She is not impressed by the structures stating "that there is little about these totem poles which is at all attractive … yet he makes miniature representations of it (the totem poles) … and sells them to tourists in the summer."[52] Edward Parkinson describes the Tlingit natives at Wrangell, which he calls Fort Wangle, as "not unlike their civilized sisters in one respect—the love of display of jewelry. Many of them are adorned with numerous rings and bracelets, which they would sell to the tourist at good prices."[53] As soon as the ship pulled into port and the gangplank was lowered, he noted that "the Indians were seen coming with their baskets and trinkets to sell to tourists at exorbitant prices."[54] Parkinson seems to put himself above his fellow travelers in his narrative as he did not purchase any curios during his travels.

Lillian Manberg Bundy, from Centralia, Illinois, took a six-week circle tour to Southeast Alaska and the Yukon in 1929. Arriving on June 18th in Ketchikan, she "walked around holding an umbrella" in reference to a typical rainy day. "Shopped at a curio shop bot (sic) baskets and a totem pole."[55] Bundy does not state which shop was visited but there were several shops in Ketchikan in 1929. The names of the curio shops are rarely mentioned in the journals. Florence Wendt published her travel account in 1935. She traveled on the S.S. *Yukon* of the Alaska Steamship Company with stops in Ketchikan, Wrangell, Petersburg, Juneau, Haines,

---

[49]Collins, p. 77.

[50]Collins, p. 77.

[51]Collins, p. 78.

[52]Collins, pp. 79–81.

[53]Parkinson, p. 99.

[54]Parkinson, p. 96.

[55]Bundy, n.p.

162 Language and Cross-Cultural Communication in Travel and Tourism

Skagway, and Sitka.[56] Chapter two of her account focused on Southeast Alaska. Her first remark about the first stop in Ketchikan was the weather. "June and September usually had the most sunshine—but since this was July, that knowledge was of little comfort."[57] She commented on her fellow travelers. "As soon as the boat docked, the town was over-run with the passengers. They swarmed to the curio stores, the trading posts ..."[58] It is possible that she visited Pruell's gift shop or the Knox Brothers, both well-known and well-advertised shops. She wondered about the totem poles and "opinion is divided now as to whether totem poles built in the present day are for the Indians or the tourists."[59] For Florence, the time allotted on shore was not enough to explore and "whoever guides a visitor around is either resident of the country or someone from the boat...just part of the Alaskan hospitality that has not yet been worn out by the presence of the wrong sort of tourist."[60]

The only detailed description of the cruise route between Wrangell and Petersburg so far is from Florence Wendt as she stayed up to watch the transit through the difficult and dangerous (to this day) Wrangell Narrows between Wrangell and Petersburg. "The green and red lights of the buoys"[61] mark the turns of the passenger boat to avoid grounding on a shallow reef. Much as today's Alaska Marine Highway ferries on this route, the early ships had two men on the bow, "alert and watchful ... stationed to drop anchor instantly in case a sudden stop in necessary."[62] As the boat emerges from the Narrows, "the boat heads for the harbor of the most colorful city along the coast," which was Petersburg, a Norwegian Settlement[63]. It is difficult to ascertain when the boats began to stop at Petersburg although photographs available from time were taken in the 1910s, and the cruise ship company literature suggests regular visits began in the 1920s.

The next stop for most ships was Juneau. Bundy and her traveling companions went to Mendenhall Glacier and she described it as a dead

---

[56]Alaska Steamship Co., n.p.

[57]Wendt, p. 11.

[58]IBID.

[59]Wendt, p. 11.

[60]Wendt, p. 13.

[61]Wendt, p. 16.

[62]Wendt, p. 17.

[63]Wendt, p. 18.

Imagined Communities: The Development of the Early Tourism Industry 163

glacier, meaning that it did not calve into the water like the tidewater Taku Glacier. There is no mention of interaction with the Auk or Taku Tlingits who often sold items near the docks. Bundy was born on Douglas Island and she "stayed up to see my birthplace. 28 years ago, I was here" as the ship departed Juneau.[64] Florence Carr Wendt gives the best detail of Juneau and its environs in her 1931 travel journal, describing the town as "framed-in residences are set in well-kept, sometimes landscaped lawns." She discusses the gravel road that was constructed to take visitors out to Mendenhall Glacier as well as to open more of the area to settlement.[65] Wendt does not discuss any curio shops in Juneau although she did remark on the fewer number of totems that in her other stops.

Mary Hayes was in Juneau on the evening of July 19 and she stated it was raining so they did not stay ashore for long. She did mention "it has numerous shops and a few good buildings" and she mentions Treadwell mines across the channel on Douglas Island.[66] At this point, Hayes no longer comments on the stops in Southeast Alaska except to say that she was not worried about touring Skagway as their "short time in Skagway as it is an uninteresting place in looks" as they boarded the White Pass and Yukon Railway for Whitehorse.[67] Collis states that Juneau "is really a mining camp, founded by James Juneau and Richard Harrisburg just 10 years ago" and "it is to-day the most important commercial point upon the entire coast.[68] She describes several curio shops in Juneau "at which may

---

[64]Wendt, p. 20.

[65]In Stroller's Weekly published on Douglas Island, a discussion of the construction of the road to the Valley is mentioned briefly. The short article is title "Glacier Road to be Scenic Triumph" and states that the progress made by "the Bureau of Roads, Department of Forestry, on the construction of the new highway, which will skirt Mendenhall Glacier and which when completed will be one of Juneau's most valuable assets from a scenic standpoint ... it being possible to reach it over the smooth Glacier Highway" (Stroller's Weekly; June 17, 1922). "Along the shores of Gastineau Channel and through the Tongass National Forest is a fine gravel highway ... it winds past lowlands of the Alaska cotton ... and blue lupine and purple fireweed brighten the meadows" (Wendt 20). She describes Mendenhall Glacier as "two miles wide and several hundred feet wide" (Wendt 20). The front of the glacier was just beyond roughly where the current visitor center is located today. It has been in retreat for over 200 years. Auk Lake, not far from Mendenhall Glacier, "a green gem set in an unbroken shoreline of spruce and cedar" is still a popular place today for visitors (Wendt 20). Her group traveled what is known as the Loop Road today, which links Mendenhall Glacier Road with the Glacier Highway through spruce and cedar.

[66]Hayes, p. 201.

[67]Hayes, p. 202.

[68]Collins, p. 156.

164   Language and Cross-Cultural Communication in Travel and Tourism

be purchased every known Equimaux (Eskimo) curio."[69] Collis visited a shop run by "Messrs. Kohler and James, who … are the successors of the Northwest Trading Company" and describes perhaps a local Tlingit selling furs stating, "the Indian receives in payment a number of blue or red tickets, which are taken by the store-keeper in exchanges for such commodities."[70] She purchased furs from this store.

Sitka was a part of early itineraries through 1906 because it was the territorial capital until it was moved to Juneau, a much better location on the Inside Passage on the way to Skagway. Travelers remarked on St. Michael's Russian Orthodox Church, which was an architectural treasure in the town. Parkinson mentions the Tlingit women with their wares for sale upon their arrival in Sitka, describing them as the "customary array of Indian squaws…with their trinkets spread out before them … and tourists soon returned to the boat with armloads of totems—sticks, small canoes, baskets, canoe-paddles, and other things too numerous to mention."[71] Parkinson described Princess Thom (Tom). He walked through the Tlingit section of town, "hunting for Princess Thom, fat, wrinkled Indian squaw has amassed a small fortune in trading among her tribes and with the white people."[72] This passage illustrates the divergent views of the curio trade. For white tourists, the curios were nick-knacks to be collected as symbols of their explorations and exploits. For Alaskan Natives, the curios were representative of the ability to adapt their traditional crafts and lifeways to a colonial capitalist system. Curios were a tool of Indigenous sovereignty. If tourists wished to consume and view Alaskan culture, they would pay for it. And in return, women like Princess Thom would rise in rank in their own society thanks to their business acumen. Parkinson's statement, while grudgingly complimentary, represents the gendered norms of the colonial paradigm toward financially successful women.

9.4 THE CURIO SHOPS

Thanks to the Gold Rush, Skagway would become the most important stop on the cruise tour of Southeast Alaska. Herman Kirmse opened his curio

---

[69]IBID

[70]Collins, p. 167.

[71]Parkinson, p. 100.

[72]Parkinson, p. 101.

Imagined Communities: The Development of the Early Tourism Industry   165

shop in 1897 and became the exclusive concessionaire for the White Pass and Yukon Railroad (built to support gold miners). He touted his store as Alaska's largest curio shop.[73] In an undated catalogue from his curio shop, Kirmse sold wholesale and retail jewelry and Native-made curios. Jewelry included brooches and crosses as well as necklaces, cuff links, bracelets, stick pins (hat pins), rings, and charms.[74] Kirmse also sold souvenir silver spoons. The Alaska silver spoon handles were cast in various Tlingit motifs.[75] Native carved wooden-handled spoons were also available. Ivory was imported from the Interior and carved into various motifs as well.[76] The miniature totems that were available were specific to each site such as the Chief Shakes Totem in Wrangell.[77] Eventually Kirmse owned a second store in Ketchikan. An advertisement in the August 9th, 1920 edition of Skagway's *The Daily Alaskan* touts him as the "pioneer jeweler of Alaska" and both locations are mentioned.[78]

Kirmse's was not the only curio shop to publish its own brochure. The Berthelson and Pruell Curio Shop in Ketchikan, run by Bert Berthelson and I.G. "Gus" Pruell, published a guide that included a brief history of Ketchikan and a short discussion of totems. In November 1921, their guide, *The Jeweler's Circular* advertised that Berthelson had sold his interest in the shop to Pruell and "is in Seattle on an indefinite vacation."[79] While the brochure gave news and information to visitors and residents in Southeast Alaska, wares for sale were advertised. By 1922, at least, they were advertising themselves as the largest curio shop in Southeast Alaska. Like many brochures, the front of the guide was decorated with a stylized totem that dominated the cover.[80]

The *Jeweler's Circular* brochure was dedicated to curios available at the Pruell shop and the experience of the visit. They will "shower curios and tell many interesting facts in connection with them."[81] Pruell's shop

---

[73] Kirmse's, p. 2.

[74] Kirmse's, p. 4.

[75] Kirmse's p. 15.

[76] Kirmse's p. 17.

[77] Kirmse's p. 19.

[78] *The Daily Alaskan,* 1920; p. 4.

[79] "The Jeweler's Circular" 1921; p. 135.

[80] "The Jeweler's Circular" 1922.

[81] "The Jeweler's Circular" 1922; p. 2.

166 Language and Cross-Cultural Communication in Travel and Tourism

carried old ivory carvings consisting of napkin rings, cribbage boards, and totem poles, a "large collection of wood and slate totems."[82] "Indian jewelry and silver spoons" were made by "Mr. Mather, a Metlakatla Indian."[83] Baskets continued to be an important part of the curio trade and Attu and spruce baskets were part of the inventory of many curio shops. Pruell also had an assortment of cedar baskets. The Attu baskets were a rare instance of an item that did not originate in Southeast Alaska. Pruell's brochure claims that the baskets were made with "Indian dyes ... and are sold at Indian prices."[84] Much like other literature, the brochure also discusses totems. As Pruell interprets, "the Natives are Darwinians to the very letter. Their belief in the origin of man from animals...on their totem poles."[85] The brochure continues to give a brief history of the totems as "recorded history, genealogy, legend, memorial, commemoration, and art."[86] On the last page is an interior of Pruell's shop in Ketchikan and it shows a variety of items including carved totems, baskets, and postcards. I. G. Pruell purchased the shop from the widow of Herman D. Kirmse in 1913. This was his second shop there and gave him a significant advantage in the curio trade by operating two large shops in routine stops on the Southeast Alaska itinerary.

Ketchikan, like Skagway and Juneau, would find ways to promote their location outside of the cruise ship literature. One way to do this was through city commercial clubs. These clubs were made up of local businessmen and women for the purpose of promoting their locations not only as place of commerce but also places to visit. In the first decade of Ketchikan's existence, two hotels were built and the Commercial Club, an early Chamber of Commerce, began to find ways to attract ever-increasing numbers of travelers to Alaska's First City, so named since it is the first city in Alaska after crossing over from Canada through the Inside Passage.[87] The late writer June Allen refers to a man named Fairbanks West as the "granddaddy of Alaska tourism" who began a travel agency that would

---

[82]IBID.

[83]IBID.

[84]IBID.

[85]IBID.

[86]"The Jeweler's Circular" 1922; p. 4.

[87]Allen, p. 2.

Imagined Communities: The Development of the Early Tourism Industry   167

promote a lucrative cruise ship line and hotel.[88] However, information on Fairbanks West had been elusive and Allen's newspaper article did not cite her sources.[89]

Lloyd MacDowell wrote the early promotional literature for the Alaska Steamship Company. He wrote several booklets including a full-color booklet called "Alaska Indian Basketry" in 1904 and another one in 1906 called "The Totem Poles of Alaska and Indian Mythology." For their time, these MacDowell booklets were enlightened small volumes on Alaska's totem poles and basketry although the language is typical for the time period. Outside of their collection value for museums, not much attention was paid to the interpretation of the poles. MacDowell states "that little is known of the real meaning of these carvings ... in Southeastern Alaska."[90] He provides some information on his definition of the "mythology of Alaska Indians" through information he received from Reverend William Duncan of Metlakahtla.[91] Since a visit to Metlakahtla was part of the Alaska Steamship Company's itinerary, the brief discussion on the heraldry served to signify all totems in general. The only other places that totems would be seen were Wrangell, Old Kasaan (an abandoned village on Prince of Wales Island), and Sitka. This pamphlet discusses the Haida and Tsimshian totems exclusively and mentions the Bear totem in Wrangell and the Kyan totem in Ketchikan. In his description, he states that miniature versions of these totems may be purchased at the curio shops in the towns where the Alaska Steamship Company ships visit and extolls the virtues of the Haida carvers. "The tourist will also be struck by the neatness as well as the oddness of the designs of the Haida carvers upon examining their work on silver and copper."[92] The booklet is completed by two pages of the descriptions of the two Alaska Steamship Company ships, the *Jefferson* and the *Dolphin* along with lists of amenities for both ships and testimonies from previous travelers on the beauty of Southeast Alaska and the comfort and convenience of the ships.

---

[88]Allen, p. 3.

[89]June Allen, "Ketchikan's Cruise Ship Industry: A Light-Hearted Look at its Origins" *Stories in the News*. Ketchikan, A. K., April 17, 2004.

[90]MacDowell, p. 3.

[91]IBID.

[92]MacDowell, p. 12.

9.5 CONCLUSION

The cruise industry dominated the tourist trade in Southeast Alaska until the outbreak of World War II which disrupted not only the flow of tourists and travel, but the use of ships and shipping routes. Yet the desire to visit the wilds of Alaska seem to have grown as the urban continues to outstrip the rural in growth in American development. And just as tourists of the 1920s, visitors to Alaska today seek out Native made products and experiences to authenticate their visit. The growth of tourism encourage visitors to uncover an American identity, one that was being written with each pamphlet and flyer. Alaskan Natives were attempting to find a place to survive and thrive through the crush of colonization. Into this matrix, the curio was born and took on special significance. It was something to be owned, displayed, and cherished. And for Alaska Natives, something to be shared. Curios became an art of cross cultural negotiation. Indigenous artisans of the 1920s were in control of their own production and sales and seemed to benefit from the growth of tourism and the development of the cruise industry in southeast Alaska. Though they had little control over the ethnocentric lens through which tourists and promoters viewed their products.

The tourist industry has long relied on commodification of experience in order to attract travelers to participate in travel experiences. But there is a discrepancy between touristic items for sale and the social reality of the people portrayed in those items.[93] More than that it is important to understand who produces these items for sale, why they are produced, and how they reinforce and reproduce the social inequalities that represent the realities of life for many Indigenous people. The search for the experience of other cultures is at the heart of the developing travel and tourism industries from the 1870s to the 1940s. The discrepancy between lived reality for Indigenous people and the imagery of Indigenous cultures and landscapes as well as the items produced for sale to tourists is a product of wider social mechanisms which have been created though the socio-historical conditions of land and culture loss suffered by Indigenous people as well as the "exoticizing" and fetishizing of a culture turned into an anachronism by the majority population. However, many tribal officials see benefits to the tourist industry as it offers opportunities for promoting and sharing

---

[93]Coronado, G. Selling Culture? Between Commoditisation and Cultural Control in Indigenous Alternative Tourism. *Pasos* 2014, *12* (1), 11–28.

Imagined Communities: The Development of the Early Tourism Industry   169

cultural traditions with outsiders and their own younger generations.[94] The question of power related concepts of rights, ownership and consent play a role in the development of the Indigenous tourist trade in Alaska. While growth of a leisure travel industry can exist without mass-produced souvenirs, the industry responded to consumer needs and wants in order to package trips and provide a cohesive experience for the leisure traveler. In effect, the use of "othering" and marketing to promote cruise travel affected the cultures of this U.S. territory based upon how much control the Indigenous peoples had over the creation and marketing of goods for the tourist trade, as well as how they were portrayed.

Today, the promotion and production of crafts for the tourist trade, as well as the interpretation of their own culture is largely in the hands of Indigenous people of Alaska themselves.[95] The Alaska Native Claims Act of 1971 created semi-autonomous corporations and villages who invested heavily in tourism, providing local control of the industry.[96] The Skagway Traditional Council is a federally recognized Indian Tribe and is the governing body for Alaska Native and American Indian activities in the Skagway area. The group governs mostly Tlingit members and events and in 1999 built a community center He?en Agunata?ani Hit (Whitecaps on the Water House) which hosts numerous cultural events for not only Tlingit, but also other closely allied tribes from Alaska to Canada. Events are also open to tourists and many leaders see these travelers as potential allies in the ongoing battle for land and resource stewardship.

Groups like Sealaska, one of the corporations created in 1971, play a vital role in cultural preservation as well as the promotion of tourism for both Indigenous and non-indigenous travelers. Since 1980, Sealaska Heritage has invited tourists to experience true indigenous Northwest culture through their events and museum/store/events center in Juneau. The Sealaska Heritage Store showcases authentic Native Art and the proceeds from the sale of the goods go directly to the producing artists. Their true purpose is to preserve, promote, and educate their populations as well as the public about the arts, language, and practices of the Tlingit, Haida, and Tsimshian.

---

[94]See oral interviews in Cerveny, p. v. Lee K. Cerveny. July 2005. *Tourism and it's Effects on Southeast Alaska Communities and Resources: Case Studies from Haines, Craig, and Hoonah Alaska.* Research Paper PNW-RP-566. USDA, Forest Service, Pacific Northwest Research Station.

[95]For an excellent look at the process of creating the Alaska Corporation model for the governance of Alaskan Tribes, see, Peter Metcalfe with Kathy Kolkhorst Ruddy. In *A Dangerous Idea: The Alaska Native Brotherhood and the Struggle for Indigenous Rights*; University of Alaska Press: Fairbanks, 2014.

[96]Cerveny, iv.

Since 1982, Sealaska Heritage has sponsored and hosted the Celebration festival. This event takes place every two years and is a platform for the exposition of local crafts, dance, and other cultural expressions. The event draws tourists from near and far as well as other indigenous groups who wish to participate in the cultural revival of Southeast Alaska. Cruise ships pull into Juneau alongside the canoes of the arriving and visiting Indigenous coastal nations. While the tourists departing the cruise ships might seem to be interlopers in this deeply meaningful event, they are welcomed in as guests who wish to learn more about the people of Alaska.

The tourist and leisure travel industry bring in much needed income and are a very large part of the economy in Alaska. The money spent by tourists is vital for many of the Indigenous people who participate in that economy and has become a method for promoting and preserving material culture. While initially goods sold to tourists were very close to items produced for practical use, over time artisans began to develop goods intended solely for promotion to tourists moving the tourist industry beyond tokenism and into inclusion and incorporation. The items produced and imagery portrayed, illustrate the fact that Alaska's Indigenous people are beyond erasure. *Painful Beauty,* as art historian Megan Smetzer calls these goods for they were necessitated by the process of colonization, yet their production for tourists spurred the retention of cultural practices.[97] For tourists, the lands they visit represent an imagined and idealized space. For those participating in or representing the tourist industry, this imagined space is their fixed habitation and their home.

KEYWORDS

- **Alaska**
- **cruise**
- **curio**
- **tourist**
- **indigenous**
- **southeast Alaska**
- **tourist trade**

---

[97]Smetzer, M. A. In *Painful Beauty: Tlingit Women, Beadwork and the Art of Resilience*; University of Washington Press: Seattle, 2021.

REFERENCES

Alaska, Atlin, and the Yukon. Alaska Steamship Company, 1926.

Alaska: Land of the Midnight Sun. Metropolitan Press: Seattle, WA, 1922.

Alaska Steamship Company. Seattle, WA, 1935.

An Old-Time Businessman Killed. The Daily Alaskan, Nov 20, 1912.

Boissevain, J. In *Coping with Tourists: European Reactions to Mass Tourism*; Berghahn Books: NY, 1996.

Bricker, K.; Donohoe, H. In *Demystifying Theories in Tourism Research*; CABI: Boston, 2015.

Bundy, L. M. *Circle Tour to Alaska in 1929*, 1929.

Bunten, A. C. More Like Ourselves: Indigenous Capitalism Through Tourism. *Am. Indian Q.* **Summer 2010**, *34*, 3.

Campbell, B. In *The Darkest Alaska: Travel and Empire Along the Insid   Passage*; University of Pennsylvania Press: Philadelphia, 2007.

Cerveny, L. K. *Tourism and it's Effects on Southeast Alaska Communities and Resources: Case Studies from Haines, Craig, and Hoonah Alaska.* Research Paper PNW-RP-566. USDA, Forest Service, Pacific Northwest Research Station, July 2005.

Collis, S. M. In *A Woman's Trip To Alaska: Being An Account Of A Voyage Through the Inland Seas of the Sitkan Archipelago in 1890*; Cassell Publishing Company: New York, 1890.

Coronado, G. Selling Culture? Between Commoditisation and Cultural Control in Indigenous Alternative Tourism. *Pasos* **2014**, *12* (1).

Cruising to Alaska. Alaska Steamship Company, 1935.

DeKadt, E. In *Tourism: Passport to Development?* Oxford University Press: Oxford, 1979.

de Laguna, F. Under Mt. St. Elias: The History and Culture of the Yakutat Tlingit. In *Smithsonian Contributions to Anthropology*, 1972; vol 1, p. 144.

Facts About Alaska 1935. Journal Printing Company: Ketchikan, AK, 1935.

Glacier Road to Be Scenic Triumph. Stroller's Weekly and Douglas Island News, July 17, 1922.

Graburn, N. In *Ethnic and Tourist Arts: Cultural Expressions from the 4th World*; University of CA Press: Berkeley, 1976.

Haycox, S. Mangusso, M. C. In *An Alaskan Anthology*; University of Washington   Press: Seattle, 2011.

Hayes, Mary Webb. Rutherford B. Hayes Library, Webb C. Hayes Collection. 1916. Unpublished journal.

Henay, C. There is Nowhere to Hide: Spirit and Heart in Afro-Indigenous Transformative Engagement. In *Cultural and Pedagogical Inquiry*; University of Alberta, Fall 2018.

Hill, L. J. Indigenous Culture: Both Malleable and Valuable. *J. Cult. Herit. Manag. Sustain. Dev.* **2011**, V*1* (2).

June, A. Ketchikan's Cruise Ship Industry: A Light-Hearted Look at its Origins. *Stories in The News*, Apr 17, 2004.

Kern, P. E. *Souvenir of Alaska*, Daily Alaskan, 1908.

Kern, E. K. In *Little Journeys to Alaska and Canada*; A. Flanagan Co.: Chicago, 1923.

Kirmse's New Shop Is Open For Business. In *Ketchikan Alaska Chronicle*, Oct 2, 1920.

Kirmse's Wholesale and Retail Jewelers. Skagway, AK. n.d.

Lenz, M. J. Material: George Heye and his Golden Rule. *American Indian Art Magazine*, Autumn 2004.

Likorish, L. J.; Jenkins, C. L. In *An Introduction to Tourism*; Routledge: London, 1997.

MacDowell, L. In *The Totem Poles of Alaska and Indian Mythology*; The Alaska Steamship Company: Seattle, WA, 1906.

Mansperger, M. C. Tourism and Cultural Change in Small-Scale Societies. *Hum. Organ.* **1995**, *54* (1), 87–94.

Meaux, J. M. In *Pursuit of Alaska: An Anthology of Travelers' Tales, 1879-1909*; University of Washington Press: Seattle, 2013.

Metcalfe, P.; Ruddy, K. K. In *A Dangerous Idea: The Alaska Native Brotherhood and the Struggle for Indigenous Rights*; University of Alaska Press: Fairbanks, 2014.

Norris, F. *A History of Skagway's Gardens*. The Alaska State Federation of Garden Clubs Convention, Skagway, Alaska; Sept 8–11, 1988.

Page, S. In *Tourism Management: Managing for Change*; Butterworth-Heinemann: Burlington, MA, 2007.

Parkinson, E. S. In *Wonderland, Or, Twelve Weeks in and Out of the United States: Brief Account of a Trip Across the Continent, Short Run Into Mexico, Ride to the Yosemite Valley, Steamer Voyage to Alaska, the Land of Glaciers, Visit to the Great Shoshone Falls and a Stage Ride Through the Yellowstone National Park*. MacCrelish & Quigley, Book and Job Printers: United States, 1894.

Pioneer Jeweler of Alaska. The Daily Alaskan, Aug 9, 1920.

Plan Protection of Alaska Native Curio Art. The Daily Alaskan; Jan 11, 1921.

Porter, R. D. In *Report on Population and Resources of Alaska and the Eleventh Census,1890*; Government Printing Office: Washington, DC, 1893.

Princess Steamships to Alaska. Schedule; Canadian Pacific: Alaska, 1929.

Resources and Scenic Attractions of Wrangell, Alaska; Sentinel Printing: Wrangell, AK, 1908.

Richards, G.; Munsters, W. In *Cultural Tourism Research Methods*; CABI Publishing: Wallingford, Oxon, GBR, 2010.

Robinson, P.; Heitmann, S.; Dieke, P. U. C. Eds.; In *Research Themes for Tourism*; CABI Publishing: Wallingford, Oxon, GBR, 2011.

Ryan, C. A History of Tourism in the English-Speaking World. In *Recreational Tourism Demand and Impacts*; Channel View Publications: Clevedon, England, 2003.

The Daily Alaska, Dec 5, 1885.

The Jeweler's Circular, 1921.

Sailing News. Alaska Steamship Company, July 2, 1939.

Schedule. Alaska Steamship Company, 1941.

Steamships. *The Daily Alaskan*, Jan 1, 1918.

Systematic Boosting. *Stroller's Weekly and Douglas Island News*, July 16, 1921.

Walton, J. K. In *Histories of Tourism: Representation, Identity and Conflict*; Channel View Publications: Clevedon, 2005.

Wendt, F. C. Florence Carr Wendt Collection, 1935–1967. Life Along the Yukon: As Seen by a Tourist in 1935.

CHAPTER 10

Multimodal Approach to Tourism Advertising Discourses

BUI THI KIM LOAN

Faculty of Foreign Languages, Van Lang University, Ho Chi Minh City, Vietnam

ABSTRACT

The paper aims to investigate tourism advertising discourses from multimodal approach. The study used theoretical frameworks including Martin and White (2005) appraisal framework, Kress and van Leeuwen (2006) visual grammar, and Bhatia (2005) move structure to explore how Vietnamese copywriters combine language and image to construct meaning for tourism advertising discourses. The appraisal framework was used to point out the realization of lexico-grammar in constructing evaluative meaning based on language for tourism advertising discourses. The theory of visual grammar also helped indicate how Vietnamese copywriters used images to create three kinds of visual meaning including representation, interaction, and composition in the tourism advertising discourses. Moreover, the theory of move structure was utilized to show how the copywriters choose different moves and steps to organize the advertising discourse to serve communicative purposes of advertisements. The data consisted of 100 tourism Vietnamese advertising posters which contained both language and image. The data were collected from the Internet and websites of travel companies from 2019 to 2021. The study used a qualitative method to describe the data from multimodal approach to see the mixture of language and image to make meaning for tourism advertisements. The

---

Language and Cross-Cultural Communication in Travel and Tourism: Strategic Adaptations.
Soumya Sankar Ghosh, Debanjali Roy, Tanmoy Putatunda, & Nilanjan Ray (Eds.)
© 2025 Apple Academic Press, Inc. Co-published with CRC Press (Taylor & Francis)

174 Language and Cross-Cultural Communication in Travel and Tourism

findings revealed that copywriters used fixed and optional moves and steps to structure the tourism advertising discourses. The moves included targeting the market, justifying the product, detailing the product service, establishing credentials, endorsement/testimonials, offering incentives, using pressure tactics, soliciting response, headlines, and slogan/logo. In addition, the findings also indicated that Vietnamese copywriters used lexis and grammar to express their evaluation for tourism advertising discourse. Moreover, the results showed that copywriters design layout of tourism advertisements to present three kinds of meaning based on visual grammar analysis such as representation, interaction, and composition. The analysis of tourism advertising discourses from the multimodal approach will be beneficial to copywriters and those who study tourism advertisements. The results of this study would also help English as a foreign language (EFL) teachers to teach tourism advertising discourses from the multimodal approach.

10.1 INTRODUCTION

Vietnamese advertising discourses have been analyzed with different theoretical frameworks. However, in Vietnam, the descriptive analysis of the tourism advertisement has not yet been researched deeply and thoroughly, and the researches have not used Systemic Functional Linguistics (SFL) framework to investigate the process of making meaning for Vietnamese advertising discourses by using both language and image. Some key terms of SFL theory that are related to genre, realization of appraisal, and visual grammar are provided in the study. The study aims to investigate tourism advertising discourses from the multimodal approach to shed light on generic structure, appraisal language, and visual grammar in the Vietnamese advertising discourses. So, to achieve the aim of the study, one research question was designed as follows:

What are the move-step structure, realization of appraisal, and visual grammar in Vietnamese advertising discourses?

10.2 LITERATURE REVIEW

10.2.1 SFLS AND ADVERTISING DISCOURSE

SFL pioneered by Halliday in 1960s is considered as a theory of language which focuses on functions of language. Halliday (1985) mentions that meaning is a choice, which means that language and other systems of semiotics are regarded as a network of choices and these choices are related together. SFL approach helps analyze and explain how meanings are created in daily communication (Eggins, 1994). Halliday also maintains that language is structured to fulfil three kinds of meanings, which are known as meta-functions at the same time. These meanings consist of ideational, interpersonal, and textual meanings. Therefore, language is regarded as a system of semiotics and a system of conventionalized codes, and structured with a range of choices. The ideational meaning shows the experience of participants who take part in the communication. Interpersonal meaning describes the social relationship of the writer and reader, and textual meaning is related to the organization of semiotic meanings to realize the ideational and interpersonal meanings. He also points out that the three meta-functions have a rapport with situational context which influences the nature and meaning of discourse. Thus, the situational context is considered as a realization of the three meta-functions.

From SFL perspective, lexico-grammar has a function of realizing register, and language is seen as a potential of making meaning for discourse. Language is analyzed according to three different strata (semantics, lexico-grammar, and phonology). Grammar is described as a system of choices but not rules. Thus, language is a meaning potential, and meaning is social-based.

10.2.2 GENRE

Martin (1992) maintains that discourse structure is seen as "schematic structure" of genre (p. 505), and genre is defined as social process which has purposed and staged orientation, thanks to register. Register includes field, mode, and tenor (Halliday, 1985). Thus, discourse structure and genre have a relationship with contextual situation.

To investigate the move-structure of Vietnamese tourism advertisements, the study used genre-based approach to analyze different moves and steps of the advertisements. The analysis of moves was introduced by Swales (1990) who analyzed the four-move model of the article's introduction. Then, this model was developed further by Kathpalia (1992) and Bhatia (1993). Although Bhatia (2004) analyzed the move structure in print advertisements, his model was used by other researchers to study "online advertisements" (Barron, 2006) and "website" commercials (Koteyko, 2009). This shows that this move-step analysis model is suitable for the analysis of this study's collected data. Kathpalia (1992) found nine-move analysis model, but Bhatia (2005, p 214) identified 10-move structure in the print advertisements. However, Bhatia (2005) did not examine the model deeply and thoroughly. Besides, there is the move of identifying the product in Kathpalia's (1992) study, but the move of evaluating the product belongs to the move of detailing the product or service in Bhatia's (2005) model. This study chose Bhatia's (2005) model of 10-move structure to investigate the move-step structure of the Vietnamese advertisements.

Bhatia (2005, pp. 213–225) points out that most print advertisements utilize some of the following "rhetorical moves" to persuade potential customers to buy products or services that are advertised.

TABLE 10.1 Move-Step Structure of Advertising Discourse (Bhatia, 2005).

Number	Move-step structure of advertising discourse
1	Targeting the market
2	Justifying the product
3	Detailing the product or service
4	Establishing credentials
5	Endorsement or testimonials
6	Offering incentives
7	Using pressure tactics
8	Soliciting response
9	Headlines
10	Slogan and logo

However, in reality in business, copywriters do not necessarily utilize all of these 10 moves in one advertisement. The copywriters will choose appropriate moves to serve communicative purposes of advertisements.

10.2.3 APPRAISAL FRAMEWORK

Martin and White (2005) appraisal framework focuses on interpersonal function of language. This theoretical framework helps identify explicit and implicit evaluation of "appraised entities" including language and images (Al-Attar, 2017, p. 103). This framework is oriented toward the appraisal of language from the writer's perspectives.

Unlike other appraisal models, Martin and White's (2005) framework is identified not only as single words but also clauses, which is known "discourse semantics" (p. 9). The appraisal framework consists of three systems such as "attitude," "graduation," and "engagement." This framework allows "inscribed and evoked appraisal" to help classify the writer's appraisal. This study only uses "graduation" which comprises "force" and "focus." Graduation depends on force and focus to convey positive and negative attitudinal meanings. Force is realized thanks to lexico-grammatical units which increase meaning and describe quantity.

10.2.4 VISUAL GRAMMAR

Visual grammar analysis model of Kress and van Leeuwen (1996, 2006) is based on SFL theory. This model indicates that, like other semiotics, images can realize the meta-functions through "representation, interaction, and composition" (Kress and van Leeuwen, 2006, p. 15). Visual grammar analysis mainly focuses on the images rather than language.

The position of placing images "vertically" or "horizontally" brings meanings for advertisements. Kress and van Leeuwen (1996, 2006) mention that three elements to identify the organization of advertisements are mainly "information value," "salience," and "framing." Different positions of the images create different meanings for advertisements. According to these two authors, the elements placed at the top of the image presents "ideal" meanings or "general" information of products, whereas the ones positioned at the bottom express "real" information and "specific" details of products. In addition, the elements of advertisements are organized horizontally. This means that the elements placed on the left are "old/given" information while the ones placed on the right convey "new" information about products. Thus, the old information tends to be general information, but the new one is specific information about advertised products. Also, the distinction between the top and bottom of images relates to metaphorical connotations of "up" and "down" in many cultures (Lakoff and Johnson, 1980).

In images, watchers are placed in a relation with figures thanks to "perspective" and "gaze" (Jones, 2012). He states that "long shots" tend to create colder relationship, whereas "close-ups" bring some feelings of being close "emotionally" and "physically." "Modality" in the images is partly realized by "realistic" level of images. He found

178 Language and Cross-Cultural Communication in Travel and Tourism

that the images in white and black in newspapers were considered to be more "realistic" than the images in bold colors that were seen in magazine commercials.

10.2.5 METHODOLOGY

The appraisal framework was used to point out the realization of lexico-grammar in constructing evaluative meaning based on language for tourism advertising discourses. The theory of visual grammar also helped indicate how copywriters used images to create three kinds of visual meaning comprising representation, interaction, and composition in the tourism advertising discourses. Moreover, the theory of move structure was utilized to show how copywriters chose different moves and steps to organize the advertising discourse to serve communicative purposes. The data consisted of 100 tourism Vietnamese advertising posters which contained both language and image. The data were collected from the Internet and websites of travel companies from 2019 to 2021. The study used a qualitative method to describe the data from the multimodal approach to see the mixture of language and image to make meaning for tourism advertisements.

10.3 RESULT

10.3.1 ANALYZING VIETNAMESE TOURISM ADVERTISING DISCOURSES FROM MULTIMODAL APPROACH

The findings revealed that copywriters used fixed and optional moves and steps to structure the tourism advertising discourses. The moves included (1) targeting the market, (2) justifying the product, (3) detailing the product service, (4) establishing credentials, (5) endorsement/testimonials, (6) offering incentives, (7) using pressure tactics, (8) soliciting response, (9) headlines, and (10) slogan/logo. In addition, the findings indicated that Vietnamese copywriters used lexis and grammar to express their evaluation in the tourism advertising discourses. Moreover, the results showed that the copywriters designed the layout of tourism advertisements to present three kinds of meanings from visual grammar analysis such as representation, interaction, and composition.

Multimodal Approach to Tourism Advertising Discourses 179

Analyzing the Figure 10.1, it can be seen that the advertisement aims at honeymooners. The Vietnamese copywriter uses the image of the couple, place of interest, and landscapes such as sea, coconut trees, or grass field to attract reader's attention to the tourism advertising discourse. The copywriter also uses language "Tuần trăng mật 6 đêm" (*6-night honeymoon*) in combination with the images to construe meaning for the advertisement. From the image and language, the copywriter would like to convey the idea that "Đảo Phú Quốc" (Phu Quoc Island) is the perfect place for any couples who plan to be on their honeymoon. That is the reason why the man and woman in the figure look at the sky seeing the words "Đảo Phú Quốc". It is concluded that Figure 10.1 mentions the subjects or customers of the tourism advertisement.

FIGURE 10.1 Targeting the market.

In Figure 10.2, the copywriter introduces a new type of tour by using the headline named "Tour Khám phá" (Exploring Tour) to catch reader's attraction. The contents of this tourism advertising discourse describe "Đà Nẵng—Beach City." Particularly, many places of interest are mentioned by using both language (Bán đảo Sơn Trà—Cù Lao Chàm—Hội An—Bà Nà) placed at the bottom and on the right side of the advertisement and using the images to help readers imagine how beautiful these places are. On the top right side of the advertisement, there is a logo of the tourist company named "HANDETOUR" and a slogan below "Your time is well spent and memorable." This slogan is one of the ways to persuade the potential tourists to take tours after they read this advertisement by using descriptive and persuasive adjective phrases such as "well spent" and "memorable." Besides, the way the copywriter employs the pronoun "your" is rather interesting. The copywriter aims at any readers who see and read the tourism advertising discourse, and this partly helps serve the communicative and social purposes of the advertisement. It can be inferred that the Figure 10.2 mentions some moves of the advertisement by using both language and image such as justifying the product or service, detailing the product or service, headline, and logo/slogan.

FIGURE 10.2   Justifying and detailing the product or service.

It can be seen that there is a logo and slogan on the top left side of the advertising discourse. The logo is considered as a signal for establishing credentials instead of using language. This logo is well-known in tourism advertisements in Vietnam, and it can be said that the advertisement is reliable for readers to believe in what the content of the advertisement is about. In addition, this advertising discourse indicates some visual meanings such as presentation and composition. This advertisement is designed horizontally, and colors separate the images describing Singapore and the language detailing the service such as the slogan "Tour Singapore cao cấp 4 sao" (Luxurious 4-star Singapore Tour), the price for the tour "6.888.000," and the tour length "3N2Đ" (3N2D) meaning 3 nights and 2 days. In this advertisement, the copywriter also uses evaluative adjective "cao cấp" (luxurious) to express one's evaluation or appraisal for the content of the commercial. In general, the Vietnamese writers cleverly make a distinction between the language and image; however, both the language and image combine together to create meanings for the whole advertisement.

FIGURE 10.3   Establishing credentials.

Figure 10.4 presents the move of endorsement/testimonials. The copywriter often uses famous people in Vietnam or abroad to realize the endorsement/testimonial move. The famous people or celebrities might be singer, MC, actor, athlete, or anyone who can influence the public and readers, and they are considered as influencers. The move of incentives is also mentioned in this tourism advertising discourse, thanks to the language such as "Siêu khuyến mãi" (Super sale), "–50%" (50% discount), and "Chỉ còn 7.500.000" (Just only 7.500.000). Therefore, the copywriter employs the words like "super sale," "just," or a semiotic "–50%" to inform readers of promotion programs and persuade them to enjoy advertised services.

182　　　Language and Cross-Cultural Communication in Travel and Tourism

FIGURE 10.4　Endorsement/testimonials and offering incentives.

It can be seen in Figure 10.5 that the moves of using pressure tactics and soliciting response are used in the advertisement. The move of using pressure tactics is realized by mentioning time limitation that readers have to book tours. The copywriter uses the time frame such as "Từ 05/07/2020 đến 30/08/2020" (From 05/07/2020 to 30/08/2020) and "Khách đặt trước tối thiểu 5 ngày, trước ngày sử dụng dịch vụ" (Book tour 5 days in advance, before the day you use service) to remind the readers of making quick decision on booking tours on time. Otherwise, they do not have chances for the tours. Additionally, the copywriter uses the move of soliciting response to provide some specific ways for the readers to contact the travelling agency. Hence, potential customers can reach the tourism service provider by calling the hotline number "0913.333.777," sending emails

Multimodal Approach to Tourism Advertising Discourses   183

to the email address of sales sector named sales.KLFtravel@cfscorp.vn, or visiting the website of company called www.klftravel.vn for further information. The interactive meaning of the image can be easily seen by looking at the man in the advertisement who is looking at the readers and smiling at them. The copywriter is trying to create a close interpersonal relationship with the readers.

FIGURE 10.5   Using pressure tactics and soliciting response.

10.3.2   THE MOVE-STEP STRUCTURE OF VIETNAMESE TOURISM ADVERTISING DISCOURSES

The findings revealed that Vietnamese tourism advertising discourses used some perspectives of SFL theory to make eye-catching and effective advertising discourses by employing both language and image. In other words, the multimodal approach facilitates the Vietnamese writers in designing these lively and persuasive advertisements. The representations of some key terms of SFL such as genre, appraisal framework, and visual grammar are synthesized in Table 10.2 to make clear how these theories are applied into making effective advertisements.

TABLE 10.2 The Move-Step Structure in Vietnamese Tourism Advertising Discourses.

Number	Moves	Steps
1	Targeting the market	Groups of customers named explicitly Nature of potential customers Usage of unique selling declaration
2	Justifying the product	Introduction of new features Improved features Additional benefits
3	Detailing the product or service	Adjectives Expressions in the form really/very/so/just, etc. + adjective Comparative, superlative Nominalization Figurative words (metaphor, personalization) Quantifiers
4	Establishing credentials	Mentioning brands and trademarks of the company Reputation Achieved awards History of the company
5	Endorsement or testimonials	Usage of celebrities and influencers in the society such as actor, singer, MC, athlete, etc. Positive comments from customers and famous people
6	Offering incentives	Promotion program Discount Free gift Voucher Free maintenance Free delivery Product trial Low interest
7	Using pressure tactics	Limited promotion time Availability of promotional products
8	Soliciting response	Providing phone/cellphone number Company website Email address Filling forms QR code Link Social network account (Facebook, Zalo, YouTube, etc.)
9	Headlines	Using one of the nine moves left
10	Slogan and logo	Short and concise language Eye-catching images

Moreover, the findings indicated that the Vietnamese copywriters flexibly used the moves. However, some eight of ten moves such as (1) targeting the market, (2) headlines, (3) slogan/logo, (4) detailing the product or service, (5) offering incentives, (6) using pressure tactics, (7) establishing credentials, and (8) soliciting response were used the most. The other two moves (e.g., justifying the product and endorsement/testimonials) were less employed, which could be explained by the fact that the copywriters prioritized important moves to convey the messages of advertisements along with the images to serve communicative purposes of the tourism advertisements.

Besides, the Vietnamese tourism advertising discourses utilized Martin and White's (2005) appraisal framework to find out the realization of appraisal in the move of detailing the product or service. The findings revealed that these move and steps were used to express the writer's appraisal or evaluation towards the products or services advertised by using adjectives, combinations of lexico-grammar, comparative/superlative expressions, figurative words, quantifiers, and intensifiers for processes or qualities.

Also, three elements including information value, salience, and framing of advertising images were exploited to make meanings for the Vietnamese advertising discourses. They arranged the language and images horizontally much more than vertically. Thus, the Vietnamese copywriters need to be flexible to organize the content of advertisements which consists of language and image appropriately to catch the reader's attention and persuade them to buy products or use services.

The findings demonstrated that the Vietnamese copywriters creatively and purposely selected colors to highlight important information of tourism advertisements. The contradiction of colors and framing helped distinguish the old/new information and ideal/real information in the advertisements as can be seen in Figures 10.6–10.8.

10.4 CONCLUSION

There are some implications from the results of analyzing the move-step structure, realization of appraisal, and visual grammar of the Vietnamese advertising discourses based on the SFL theory, particularly multimodal approach. First, it depends on the social purpose of tourism advertisements

FIGURE 10.6 Visual grammar.

FIGURE 10.7 Contradiction of colors.

Multimodal Approach to Tourism Advertising Discourses 187

FIGURE 10.8   Ideal and real information.

so that Vietnamese copywriters can flexibly choose appropriate moves and steps to make successful and effective advertisements. They do not necessarily follow the fixed order of move-step structure; however, they should pay attention to psychological factors of Vietnamese and English potential customers to select the best moves and steps for each advertising discourse. Second, Vietnamese copywriters should use the most common moves and steps to describe products or services, and to persuade the readers to buy and use the services. They should be careful to utilize nominalization or grammatical metaphor carefully in Vietnamese advertisements since there is an argument that there is no nominalization in Vietnamese due to different language types. Third, Vietnamese copywriters should apply Martin and White (2005) appraisal framework into detailing the products or services by using sources of lexico-grammar to make meaning for advertisements. Vietnamese copywriters can use graduation system including force and focus to express their appraisal for advertised products and services. Due to this appraisal framework, the readers can make sense of the contents of advertisements, and then they will do some actions such as searching the products or services and buying them. Fourth, Vietnamese copywriters need to arrange language and image in the advertisements horizontally or vertically according to different cultures. The images can carry three kinds of meanings such as representation, interaction and composition, and the Vietnamese copywriters should combine both language and image to create eye-catching and effective advertisements. Finally, the results of the study will provide useful knowledge for the teachers who have faced challenges with teaching Vietnamese and English ADs. The analysis of tourism advertising discourses from the multimodal approach will also be beneficial to copywriters and those who study tourism advertisements. The results of this study will help EFL teachers to teach tourism advertising discourses from the multimodal approach more communicatively and functionally.

KEYWORDS

- **multimodal approach**
- **tourism**
- **appraisal framework**
- **move-step structure**
- **visual grammar**

REFERENCES

Al-Attar, M. M. H. A Multimodal Analysis of Print and Online Promotional Discourse in the UK. Ph.D. Dissertation, University of Leicester, 2017.

Barron, A. Understanding Spam: A Macro-Textual Analysis. *J. Pragmatics*, **2006**, *38* (2006), 880–904.

Bhatia, V. K. In *Analysing Genre: Language use in Professional Settings*; Longman, 1993.

Bhatia, V. K. In *Worlds of Written: A Genre-Based View*; Continuum, 2004.

Bhatia, V. K. Generic Patterns in Promotional Discourse. In*Persuasion Across Genres: A Linguistic Approach*; Halmari, H., Virtanen, T., Ed.; John Benjamins, 2005; pp 213–225.

Eggins, S. In *An Introduction to Systemic Functional Linguistics*; Continuum Wellington House, 1994.

Gee, J. P. In *How to do Discourse Analysis - A Toolkit*; Routledge, 2011.

Halliday, M. A. K. In *An Introduction to Functional Grammar*; Edward Arnold, 1985.

Kathpalia, S. S. A Genre Analysis of Promotional Texts. Ph.D. Dissertation, National University of Singapore, 1992.

Koteyko, N. I am a Very Happy, Lucky Lady, and I am Full of Vitality! Analysis of Promotional Strategies on the Websites of Probiotic Yogurt Producers. *Crit. Discourse Stud.* **2009,** *6* (2), 112–125.

Kress, G.; van Leeuwen, T. In *Reading Images: The Grammar of Visual Design*; Routledge, 1996.

Kress, G.; van Leeuwen, T. In Reading Images: The Grammar of Visual Design, 2nd Ed.; Routledge, 2006.

Lakoff, G.; Johnson, M. In *Metaphors we live by*; University of Chicago Press, 1980.

Martin, J. R. In *English Text: System and Structure*; John Benjamins, 1992.

Martin, J. R.; White, P. R. R. In *The Language of Evaluation: Appraisal in English*; Palgrave Macmillan, 2005.

Swales, J. M. In *Genre Analysis: English in Academic and Research Settings*; Cambridge University Press, 1990.

CHAPTER 11

The Saga of Kochi: Cultural and Heritage Tourism Overview

NAVNEET MUNOTH[1], LINSON THOMAS[2], and SHUBHAM GEHLOT[3]

[1]*Department of Architecture and Planning, Maulana Azad National Institute of Technology, Bhopal, India*

[2]*Urban Planner, Kannur, India*

[3]*Data Analyst and Urban Planner, Indore, India*

ABSTRACT

The rich culture and heritage of a settlement reflect its unique history and development over the course of time. Kochi, a metropolitan city on the Malabar Coast of India, is part of Ernakulam district in the State of Kerala. It has grown from a small coastal settlement to economic, financial, and tourism center of the State of Kerala. Before the colonial era, it had its identity as the spice trading hub; trade of goods took place through its coasts, majorly with the Arabian Peninsula. During the imperial regime, Kochi got heavily influenced by Portuguese, Dutch, and British cultures; the blend of different cultures resulted in Kochi's unique cultural development and provided it with a distinct identity. After independence, Kochi underwent tremendous development and demographic changes; it became a metropolitan port city with immense maritime and economic importance and became a logistics transport center. Exponentially growing population accompanied by urban and infrastructural demands, and haphazard growth of the city has changed its image; development activities badly affected its rich heritage and caused detrimental effects on the environment

---

Language and Cross-Cultural Communication in Travel and Tourism: Strategic Adaptations.
Soumya Sankar Ghosh, Debanjali Roy, Tanmoy Putatunda, & Nilanjan Ray (Eds.)
© 2025 Apple Academic Press, Inc. Co-published with CRC Press (Taylor & Francis)

and coastal ecosystem. The chapter endeavors to trace the chronological development of its distinct culture and heritage, and the changes in the urban landscape of Kochi.

11.1 INTRODUCTION

The world is urbanizing at an unprecedented scale after the year 2000—there were 371 cities with 1 million inhabitants or more worldwide; it is expected that the number of million plus cities will rise to 706 by 2030 (Population Division, 2018). India is also experiencing the urbanization phenomenon; its urban population has increased to 31.16% in 2011 from 10.86% in 1901 and it is expected that contribution of cities to India's GDP will rise to 70% by 2030 (Sankhe et al., 2010). Cities will become growth engines of economy; metropolitan regions and tier-2 cities are going to play a vital role in it.

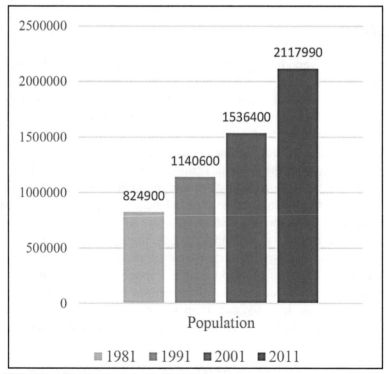

FIGURE 11.1 Rise in population in Kochi Metropolitan Area.

Kochi is one of the fastest developing tier-2 cities in India. Apart from being a commercial, industrial, educational, health, and entertainment hub of the state, it is a significant tourist destination. The evolution of Kochi as a prime city on the Malabar coast is closely associated with its political and administrative history (Department of Town and Country Planning, 2010). At the beginning of the 20th century, Kochi had a population of 0.06 million which became 0.6 million in 2011; the density of the city as per the census 2011 was 6350 persons/km². The significant change in Kochi's development came from the British intention to make the city an export–import hub during 1921–1931. Post-Independence Kochi experienced a demographic explosion triggered due to in-migration and industrial development in 1971–1981, which led to the expansion of the city to accommodate the increased population. The exponentially rising population changed the city's built fabric, which led to the conversion of available open spaces to residential and commercial buildings.

Kochi is one of the cities with the least per capita availability of open spaces. The urban expansion necessitated the demarcation of Kochi Metropolitan Area; population-wise, it is the 17th largest urban agglomeration in India and the largest urban agglomeration in Kerala (Department of Town and Country Planning, 2011). Further, it covers 440 km² having a 21.1 million population with a growth rate of 37.9% (2001–2011). Kochi Municipal Corporation comprises Kochi Mainland, Fort Kochi, Mattancherry, and Willingdon Island, covering a 94.88 km² area divided into 74 wards. Kochi city had undergone various changes in its urban form; the unique character of the city open ups an avenue to study the changes and track down its transformation. The profile of the city will assist in understanding its unique evolution from being a small harbor settlement to one of the most influencing urban centers in the region.

11.2 METHODOLOGY

The chapter uses secondary data from government sources. The data comprises of thematic maps, reports, official documents, and newspapers. Various studies, research articles, government reports, news articles, and reports from questions regarding the state at Rajya Sabha (Upper House), Lok Sabha (Lower house), and state Legislative Assembly were also analyzed. A review of planning strategies and policies that helped in Kochi city's evolution was also analyzed. The chapter explains the development of Kochi and the challenges it confronts. The chapter further discusses the significant issues and gives a path for future developments of Kochi city.

11.3 STUDY AREA

With its surrounding locations, the city provides picturesque views—Islands, Backwaters, Canals, Coconut farms, beaches, etc. A historically significant destination such as Fort Kochi and Mattancherry attracts tourists from worldwide. Kochi hosted 980 tourists per day in 2017 (Department of Tourism, 2017). The history of Kochi dates back to the 12th century when the Cochin State was formed after disintegrating from the Kulsekhra Kingdom in 1102. The strategic location of Cochin became the reason for battles between different Kingdoms. Kochi embarked on its journey to becoming a world-class port city in the 14th century when a great flood in 1341 changed the course of river Periyar and destroyed the Mucciri/Muziris Pattanam (present-day Kodungallur), which allowed Kochi to prove its significance and to develop itself as one of the commercial centers of Kerala (Census of India, 2011; Padmanabhan, 2011; Trias and Gommans, n.d.).

11.3.1 LOCATION AND GEOGRAPHY

In Malayalam, "Kochi" means "Small Lagoon" (Tourism Cochin, 2019); it is a beautiful archipelago (Action Plan for Greater, Kochi Area, 2001; Kochi Municipal Corporation, 2006) and a port city on the southwest coast of India, lying in the central region of the State Kerala. It is located between 76° 14′ and 76° 21′ East longitude, and 9° 52′ and 10° 1′ North Latitude, the city lies in the low land region of the state and is barely 2 m above sea level. The district is bounded on the north by Thrissur District, on the east by Idukki district; and the south by Kottayam and Alappuzha (Aleppey) districts. The eastern frontiers of Kochi have laterite capped low hills; many streams originate from these hills and run down to drain into the backwaters making a complex network whereas on the western side Kochi has coastal plains (Doctor of Philosophy, 2012). The serene beauty of the Kochi estuary is a natural gift to the city, being located between the Western Ghats and the Arabian Sea (Local Self Government Department, 2005) in

The Saga of Kochi: Cultural and Heritage Tourism Overview

the western coastal plains (Fig. 11.2). The city with its surrounding locations provides picturesque views—islands, backwaters, canals, coconut farms, beaches, etc., with historically important destinations such as Fort Kochi and Mattancherry attracting tourists from worldwide.

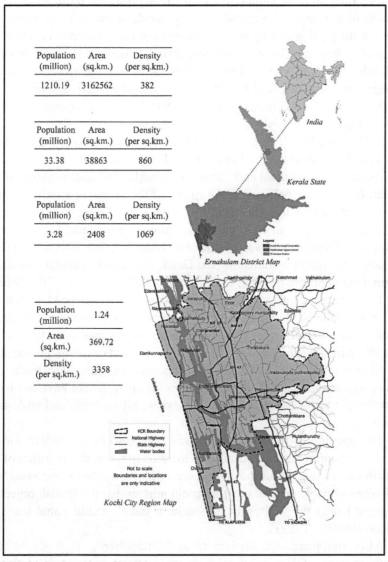

FIGURE 11.2 Location of Kochi.

11.3.2 TOURISM AND TRADE INDUCED CITY'S GROWTH

The trade and commerce in Kochi increased after the formation of Kochi harbor in 1341 resulting in the shifting of merchants and traders from Muziris to Kochi port. Kochi port gained prominence quickly, which made the then ruler shift the capital to Kochi which prompted momentum to the growth of the town. From 16th century, Kochi witnessed rapid changes through the trading and colonizing attempts of European powers (Department of Town and Country Planning, 2010). When the Portuguese came to Kochi first in 1500, they founded Fort Kochi, established factories and warehouses, schools, and hospitals, and extended their domain on the political and religious fronts (Prakash, 2001). In the 16th century, the Jews moved to Sanda Cochin from Kodungallur.

In 1663, the Dutch invaded Kochi resulting in the fall of the Portuguese (Trias and Gommans, n.d.). For a hundred years, therefore, Kochi became the center of the political and commercial battle. The interventions of the Dutch East India Company in the ruling of the kingdom continued until the second half of the 18th century. During the 1770s, the Travancore Rajas came into prominence and the kingdom of Kochi had to enter into a treaty with them. Kochi was invaded by Hyder Ali, then Mysore king and his son Tipu Sultan by defeating the Dutch East India Company. With the death of Tipu Sultan in 1799, Kochi came under the control of the British.

The erstwhile Cochin State had settlements confined to the Mattancherry and Fort Kochi area with few small-sized settlements scattered along its Northern and Eastern sides (Fig. 11.3). When Suez Canal came to function in 1869, which is going to become one of the busiest routes in the maritime world (Sir Robert Bristow, 1959); it increased the east–west traffic and trade, which necessitated the construction of ports. Kochi have not had a port then; it had a natural harbor from where all business and trade took place.

The opening of the Suez Canal was proved to be a windfall for the Kochi; its natural harbor and suitable location grabbed the attention of the British Government. But the natural site of Kochi needed some assistance to make a suitable environment for a big port so that the British objective to make Kochi the "Gateway of Southern India" would come true (Sir Robert Bristow, 1959).

After Independence settlement at Mattancherry, Fort Kochi, and southern side of Vypin Island became densely populated inhabiting

166,068 people. The city has started growing toward the Mainland (Eastern Side), and few settlements sprung up around its periphery (Fig. 11.5).

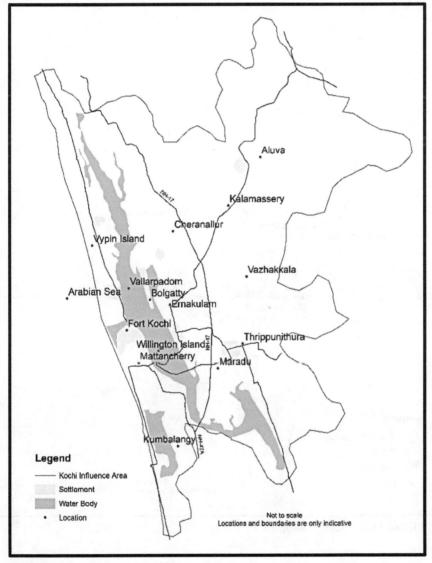

FIGURE 11.3 Settlement in 1891.

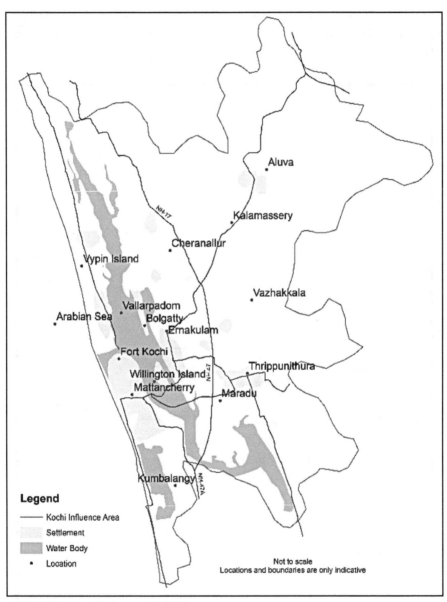

FIGURE 11.4 Settlements in 1921.

The Saga of Kochi: Cultural and Heritage Tourism Overview

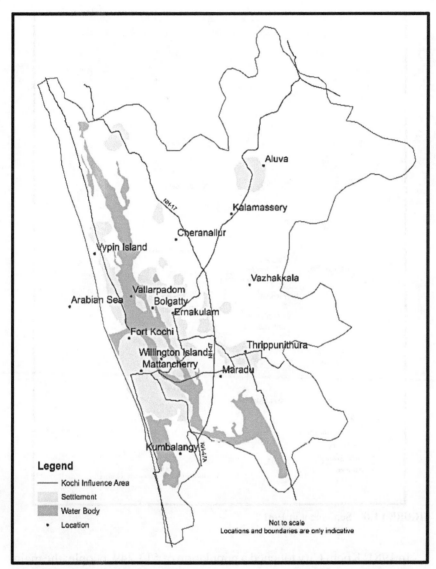

FIGURE 11.5 Settlements in 1947.

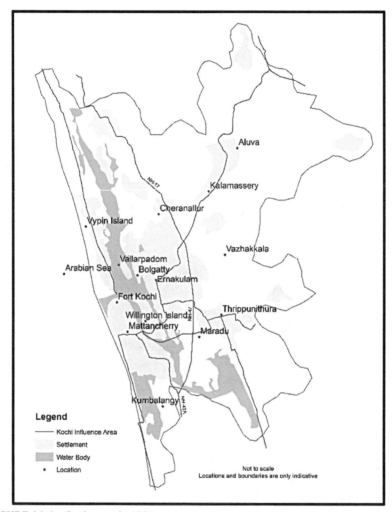

FIGURE 11.6 Settlement in 1981.

In 1981, Kochi City touched a population of 513,249, people; the mainland area of Kochi developed vigorously as the commerce and industry sectors in the city started flourishing. Mattancherry, Fort Kochi, and Vypin Island didn't undergo much change; on the contrary, the mainland settlement expanded to accommodate the rising population on the eastern side; earlier settlements had grown and transformed into small towns, which became the source of the floating population to the Kochi City (Fig. 11.6).

The Saga of Kochi: Cultural and Heritage Tourism Overview

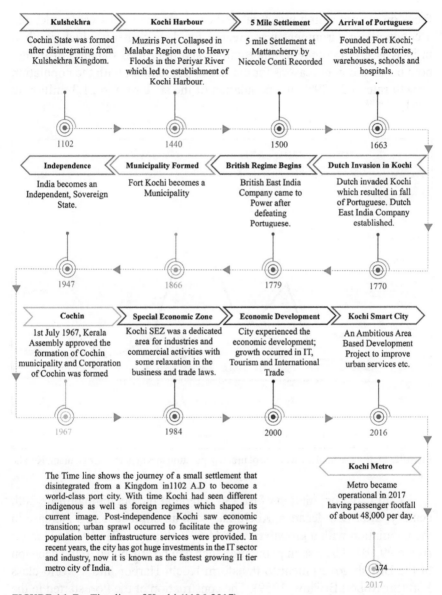

FIGURE 11-7 Timeline of Kochi (1106-2017)

11.3.3 URBAN POPULATION AND GROWTH RATE

In 1901, India had a population of 238.3 million people and the share of Kerala's total population was 2.6% with 6.3 million. After independence, focus on strengthening industrial and economic activities was made, and a conducive environment was created for development; which resulted in increased employment opportunities and improved standard of living; both became potent reasons for the rise in population; with the population growth rate of 26.29%, the population of the state rose to 21.3 million in 1971.

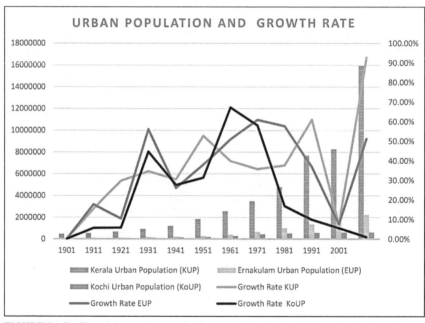

FIGURE 11.8 Decadal growth rate of urban population in Kochi city, Ernakulam, Kerala.

Kochi had a population of 61,236 people at the beginning of the 20th century; after two decades, Kochi had seen an unprecedented increase in the population with a growth rate of 44.69% during 1921–1931 making it rise to 99,101. The rise in population then can be attributed to the decision of the British government to transform Kochi Harbor into a world-class Port (Sir Robert Bristow, 1959). The construction of the Port started in the latter half of the decade; this led to a large influx of migrants as laborers, engineers, etc. The decade 1941–1951 was the most important phase in Indian History; the struggle against the British Government became intense; this phase had not seen any development.

Post-Independence Kochi city has seen an extreme demographic transition; with a growth rate of 67.23%, the population of the city became 277,723 which contributed 10.8% to the total urban population of Kerala. The old harbor city, which is transformed into most feasible urban area for investment became a potent factor for the economic development, as it was anticipated that a good port would stimulate industrial activity (Sir Robert Bristow, 1959). Kochi started its run to become a leading port city with huge financial and economic importance.

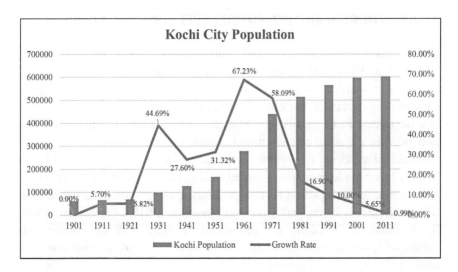

FIGURE 11.9 Kochi city population and decadal growth rate.

The decline of 9.14% has been seen in the population growth rate during the decade 1961–1971 since then; the growth rate in Kochi City dropped rapidly, which shows that city is reaching its limits and also demand land in the core city has led to price hike which will further lead to the expansion of the city. Population of Kochi city was 564,589 (0.56 million) in 1991, which rose to 596,473 (0.59 million) in 2001 with the decadal growth rate of 5.49.

Municipalities around Kochi experienced an increase in the population with a high growth rate; Kalamassery and Thrikkakkara Municipality have the highest growth rate 2001–2011, both the municipalities got industrial development and large-scale IT developments are also proposed. Kochi Municipal Corporation Area consists of around 51.2% of the KCR population with an area share of 25.6%, which indicates the varying population density in the Kochi city region with Kochi, Tripunithura, Thrikkakkara, Kalamassery, and Eloor as the densely populated settlements. During 2001–2011, Kochi Urban Agglomeration experienced a growth rate of 37.9%, whereas Kochi City had a growth rate of 0.01%, which deduce/induce that the city has reached its development capacity (Goswami et al., n.d.).

11.3.4 TOURISM AND ECONOMIC DEVELOPMENT

Ernakulam district is the economic hub of Kerala and Kochi is its nerve center of it. Historical factors, geographic factors, location, availability of natural resources, growth of service sectors, etc., highly influenced the economic activities in the city and region. The economic base of Kochi can be traced from its historic Mattancherry Market Town, which has now become a typical oriental market town, various commercial activities also take place along its waterfront. Since its inception, the Mattancherry Market has been known globally and established its presence as a cosmopolitan market; enthralling tourists and business persons from worldwide (Department of Town and Country Planning, 2010). Centuries before Kochi became the center of commercial activity with the import and export of spices to various lands. The Arabian Sea and the port played a major role in the development of Kochi city.

Kochi city provides immense employment opportunities attracting the floating population from the distance of up to 25 km^2; more than 50% of the total trips are made by commuters for employment purposes only (George, 2016; Zeba Aziz, 2018). The foreign exchange made the city famous; leading to various foreign invasions as a result. The commercial activity in the city developed with the trade of agricultural products later it moved to industrial, IT, entertainment, tourism, etc., and various other sectors made the city a commercial center of the state. A sudden economic change and development of Kochi started due to the emergence of trade and commerce due to the formation of the Kochi port in 1341. The scale of trade and commercial activities and indigenous spices attracted the Europeans. After the arrival of the British Kochi experienced drastic economic development due to the construction of railway, roads, bridges, and man-made Islands Kochi improved the inland and foreign connectivity.

The contribution of Kochi city to the state's GDP is 14.47% with the manufacturing and construction industry contributing 37.01 and 20.03% is contributed by trade, tourism, and hospitality together (George, 2016). The economy of Kochi consists of various sectors, such as fishing harbors and export of marine products, tourism and heritage, foreign exchange, Ernakulam Chamber of Commerce, Chamber of Commerce and Industry, Cochin Stock Exchange, various industries, such as, Vyppin LNG Terminal, Vallarpadam Transshipment container, Cochin shipyard, Oil refining, FACT, etc. Kochi has three Special Economic Zones, viz., Electronic Part at Kalamassery, Cochin SEZ, and a port-based SEZ which are construed to provide an economic boost to the city; also a SEZ is proposed at Kakkanad which will further increase its potential. Various MNCs and firms, and a huge population testify the economic potential of the city; IT and ITES services got accolades for Kochi naming it as the second-most attractive city in India NASSCOM (Government of Kerala, 2019).

11.3.5 TOURISM AND HERITAGE

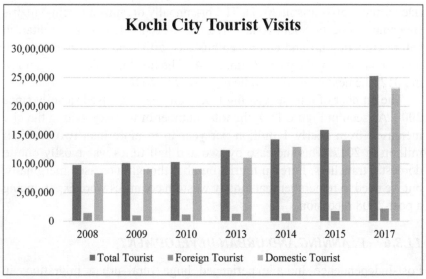

FIGURE 11.10 Rise in tourist visits in Kochi city.

Kochi is renowned for its cultural heritage and natural beauty; it attracts considerable tourist population. In 2017, total 2,525,123 tourist visited Kochi City; it received the highest domestic and foreign tourists' footfalls in the state (Department of Tourism, 2017).

Serene backwaters of Kochi, which is the cluster of islands in the Vembanad Lake, viz. Bolgatty, Vypeen, Willingdon, Gundu, and Vallarpadam Island, Fort Kochi, Santa Cruz Basilica, Mattancherry Dutch Palace, Jew Synagogue, Marine Drive, Museums, and Mangalvanam Bird Sanctuary are prime tourist destinations (Kerala Tourism, 2019; Department of Tourism, 2016). Central Kerala[1] contributed 49.4% to the total foreign tourist population in Kerala with Ernakulum's foreign tourist population to the total was about 41.58% (Department of Tourism, 2017). Tourism is the major source of income for the district and state economy; revenue generated from foreign exchange in Ernakulam district is about 3489.24 crores while state received 8392.11 crore through foreign tourism (Department of Tourism, 2017). The potential of tourism in Kochi should be harnessed by improving the infrastructure services. The rich cultural heritage of Kochi should be revived; Tourism department has undertaken projects like Major Spice Route Project and Muziris Heritage Tourism Project, which will help in promoting the culture and conserving unique the unique heritage (Department of Tourism, 2017). The bi-annually organized Kochi-Muziris Biennale is one of South Asia's five mega-events. It is cosmopolitan in nature attracting art and culture enthusiasts; artist and visitors from each nook and corner of the world (Jain, 2014). The first biennale was organized in 2012; the next biennale will be organized in 2022.

The number of tourists visiting Kochi city has increased steadily since 2008. As seen in Figure 11.4, the total number of tourists visiting the city increased from nearly 1 million per year in to more than two and half million in 2017. This increase by two and half times was mostly due to domestic travellers. Foreign tourist number though increased marginally, but we need to remember that mostly western countries were experiencing a post-2008 recession.

11.3.6 PLANNING AND URBAN DEVELOPMENT

Post-Independence India experienced huge rural–urban migration; in just 50 decades (1961–2011), it added 298.16 million people to its urban

---

[1]Central Kerala region comprises of Kottayam, Idukki, Ernakulam, and Thrissur Districts.

population. The degree of urbanization in Kerala was 47.72% in 2011, which was about 26% in 2001. Kerala got ninth rank in terms of the level of urbanization among the States of India; also Ernakulam has 68.1% of its population living in urban areas and received the title of the most urbanized district of Kerala. In 1970, the total urban population of the district was about 70%, but contemporary Kochi shares 27% urban population with Ernakulum which shows that in the past years many urban settlements grew up in the district. Nonetheless, Kochi is the fastest-growing urban center in the city having a huge economic potential (Express News Service, 2108). Kochi metro project started in 2012; it is 25.612 km long connecting Alwaye-Petta running from South-West to the North direction. The project envisaged providing better public transport services in high-density areas running parallel to the congested roads (Delhi Metro Rail Corporation Ltd, 2011). The fast-emerging city is also envisioned to become a global city by 2030 which will attract investment, boost its port-related activities, new industries will be set up, IT hubs, and tourism and trade development will occur. The concept of a Global city is incorporated in various plans so that any development which will take place helps in moving a step ahead to make Kochi a Global city.

FIGURE 11.11 Street in Fort Kochi Area, 1960.
Image courtesy: Peter Forster. https://creativecommons.org/licenses/by-sa/2.0

FIGURE 11.12 Marine Drive, Kochi.

FIGURE 11.13 Streets in Fort Kochi.

11.3.7 SMART CITY PLANNING AND DEVELOPMENT

In 2015, the newly elected Federal Government came up with an idea to make India Cities "Smart"; however, there is no particular definition for Smart City; the cities are given the liberty to demonstrate their smartness by adopting new innovative techniques for better governance and management of the cities. Initially, 99 cities were selected for the Smart City Mission which will focus on improving transportation, energy and ecology, water and sanitation, housing, and economy; these five development heads have got 80% of the Smart City Mission Budget (Anand et al., 2018).

The mission endeavors to develop an urban ecosystem that is represented by the four pillars of comprehensive development—institutional, physical, social, and economic infrastructures while ensuring sustainable and inclusive development (Ministry of Urban Development, 2015).

The Smart City Mission of Kochi has the vision to make the city an economic growth engine having efficient urban services, sustainable growth, and ease of living. The vision has four themes in which specific goals are set to achieve the desired aim. The major issue that Indian Cities are grappling with is a reduction in mobility due to congestion on roads and high ownership of private vehicles; Kochi city has witnessed 12% vehicular growth annually the second-highest growth rate in the vehicular population among million-plus cities in India (Government of India, n.d.).

The Smart City Mission is focused on improving the accessibility in the city by providing Public Transport services with increasing its shares to at least 60% in the coming 10 years, the congestion on roads will be reduced further by promoting eco-friendly mobility by pedestrianizing and providing NMT infrastructure, the public bike-sharing system is also proposed in the Smart City Area and metro corridors; the eco-friendly transportation services will help to curb the vehicular emission, improve the air quality of the city while providing the last mile connectivity. Efficient waste collection and segregation will help to make the city clean green, safe, and healthy; increasing the coverage of public toilets services will assist in achieving the aim of a 100% defecation-free city. Management of the population can be eased by affordable housing, 24 × 7 clean water supply, and an efficient sewer system. As the level of services will improve and the standard of living will rise with the efficient implementation of planning proposals under the smart city mission, it may act as a potent factor for in-migration, so the planning efforts should be reasonable in the future growth.

11.3.8 TRANSPORTATION

The unique location of Kochi makes it suitable for all modes of transportation providing it with better connectivity making a conducive environment for business and commercial activities; the city is often attributed as the nerve center of the state. The extensive road network with three national highways (NH-17, NH-47, and NH-49) makes Kochi an easily accessible city from any part of the region. Kochi city has 614 km of total road length having a road density of 1.03 km/1000 population and 6.47 km/km^2 of surface area (Department of Town and Country Planning, 2010).

Kochi CBD area poses five major arterial roads; the north–south corridor is connected by MG Road, Shanmugam Road, Chittoor Road, and the east–west corridor, viz., Banerji road and S.A road (City Mobility Plan-Kochi, 2007).These arterial roads link with the other parts of the city through sub-arterial and collector roads forming a broken gridiron pattern (Department of Town and Country Planning, n.d.). Kochi faces frequent traffic snarls mainly at the railway crossing area; the Thiruvananthapuram–Thrissur railway line bifurcates the city passing through its center. Closure of gates at crossings restricts the traffic flow, which results in an increased number of vehicles on the route, causing traffic jams. Kochi International Airport is located at Nedumbassery about 28 km away from the city. The proximity of airport to the three national highways provides it with efficient connectivity; also, the airport has a special route that connects it with seaport named Airport-Seaport road, which runs parallel to NH 47. Kochi Lagoons has 1100 km of waterways available; however, only 40 km is suitable for navigation; the two jetty stations are available in the city which offers public transport facilities but they are also facing a decline in trips (Joseph, 2012).

11.3.9   URBAN INFRASTRUCTURE

11.3.9.1   WATER SUPPLY

Kochi got its first filtered water supply system installed in 1914 by developing Chowara waterworks on the Periyar river; new waterworks at Aluva (located 20 km from Kochi) were constructed in 1965 with 48 MLD capacity having Periyar river as its source later the capacity is increased to 70 MLD in 1969 (Kochi Municipal Corporation, 2006). The World Bank assisted in installing a new treatment of 70 MLD in 1993; under a HUDCO project, a 35 MLD treatment plant was also installed at Aluva. The total capacity of the Aluva water treatment Plant became 225 MLD. Aluva Water Station supplies water to Kochi Corporation, 2 municipalities, and 16 panchayats covering 352.98 km^2 area; whereas, the current demand in the service is about 360 MLD. Under City Development, 2006 proposals have been made to install two water treatment plants; one at Maradu (100 MLD) with Muvat-tupuzha river, and one at Kalamassery (285 MLD) from Periyar river as its source to meet the ultimate demand in the year 2036 (Department

The Saga of Kochi: Cultural and Heritage Tourism Overview 211

of Town and Country Planning, n.d.). Coastal Areas at Kochi have a high-density population; to fulfill the water demand in coastal areas, desalination plant needs to be installed. Western parts of Kochi face water scarcity the per capita water availability is around 25–30 LPCD while the Ernakulam Region has water availability to be around 90 LPCD. City also faces scarcity of clean drinking water due to the presence of high pollution levels in water resources (Kerala Perspective Plan 2030, n.d.).

11.3.9.2 SEWAGE AND SANITATION

In 1970, Kochi got its first sewerage system commissioned (Kochi Municipal Corporation, 2006); the corporation area is divided into four zones with the target to cover the entire 94.88 km^2 area but the existing sewerage system only covers 5% corporation area (2.5 km^2 in the General Hospital area and 1.50 km^2 in Gandhi area) (G. and G. of India). The city has a sewage treatment plant at Elamkulam of capacity 4.50 MLD having an activated sludge treatment system (Department of Town and Country Planning, 2010). The flat terrain, high groundwater level, and barely 2 m above the mean sea level position of Kochi make it a challenge to provide a sewage system. In such a case, on-the-plot disposal system is used in the city; as this has low-cost and least-maintenance expenditure, it is incorporated. Kochi city region generates 670 tons of solid waste per day with 300 tons of contribution from Kochi Municipal Corporation area having per capita waste generation of 482 g/day/head (Department of Town and Country Planning, 2010; Hridya et al., 2016). As per CDP, only about 50% of total waste is collected and transported; the left-over waste is either incinerated or finds its way to open plots, drains, water bodies, and sea. The Solid Waste collected from Kochi Municipal area and Panchayat area is treated at Brahmapuram Treatment Plant of capacity 250 tons/day, which is insufficient—and is not even sufficient for the Municipal Corporation area (Kumar et al., 2015). The lack of solid waste and waste water treatment services is affecting the environment of the city making it a highly polluted settlement; the industrial waste further aggravates the situation by polluting air and water resources.

212 Language and Cross-Cultural Communication in Travel and Tourism

TABLE 11.1 Urban Services and Authorities in Kochi.

Urban services	Planning	Implementation	Operation and maintenance	Tariff fixation
Water supply	Kerala Water Authority	KWA, Cochin Port Trust (for port areas)	KWA, Cochin Port Trust (for port areas)	KWA
Solid waste management	Corporation of Kochi and other local bodies	Corporation of Kochi (Health Department) and other local bodies	Corporation of Kochi (Health Department) and other Local Bodies, private sector initiatives like CREDAI, Clean City Kochi etc.	Corporation of Kochi
Storm water drainage	Corporation of Kochi	Corporation of Kochi, Engineering Department (Construction), Kerala Public Works Department	Corporation of Kochi and Kerala Public Works Department	Not applicable
Sewage and sanitation	Kerala Water Authority, Corporation of Kochi	Kerala Water Authority, Corporation of Kochi ,Cochin Port Trust (for Willingdon Island) and Other Local Bodies	Kerala Water Authority, Cochin Port Trust (for Willingdon Island) and Other local bodies.	Kerala Water Authority

Source: G. and G. of India (City Sanitation Plan for Kochi).

TABLE 11.2 Quantity of Waste Generated in Kochi City Region.

S. No.	Type of waste	Quantity in MT/day	% of total
1.	Household domestic	330	55
2.	Hotels/eateries	36	6
3.	Markets/slaughter houses	30	5
4.	Shops & commercial establishment	90	15
5.	Building construction waste	30	5
6.	Garden trimmings/plantation/tree cuttings	24	4
7.	Institutional waste	30	5
8.	Industrial waste	18	3
9.	Hospital/clinics	12	2
	Total waste generated/day	600	100
	Waste collected/day	40	
	Collection efficiency	40%	

Source: Department of Town and Country Planning (2010) (Draft Development Plan for Kochi, 2031).

The Saga of Kochi: Cultural and Heritage Tourism Overview 213

11.3.9.3 SLUMS AND HOUSING IN KOCHI CITY

The unemployment rate is much higher in the slums of Kochi; according to Kochi Corporation, one-fifth of the unemployed lived in the slums in 1996 (Prakash, 2001). Most of the people living in slums belong to the lowest social and economic strata. Severe unemployment is forcing the youth and educated labor force to migrate to other parts of India and abroad for employment. Another major issue was sanitation, only 57% of the total slum population had access to toilets in 1996 (Government of Kerala Local Self Government Department, 2005). The average density of the slum is very high so that only 30% of households have space to build septic tank (Government of Kerala Local Self Government Department, 2005). Shortage of affordable housing in the Kochi city area resulted in the rise of slums. The high land cost affects the affordability of houses. According to 2001 census data, 32% of BPL population is living in slum areas/colonies.

TABLE 11.3 Changes in Share of Slum Population.

Sl. no.	Year	Total slum population	Total city population	Percentage of population living in slums to the total city population	Share of total population living in slums to the population of Kerala	Share of total population living in slums to the total population of India
1	1996	64.348	584,982	11.00%	–	–
2	2001	7873	596,473	1.32%	2.10%	23.47%
3	2011	5178	602,046	0.86%	1.27%	22.44%

Source: Census of India (2011) Kochi Census Data Reports.

The appropriate implementation of various schemes and projects developed by the Central and State government by the ULB helped to reduce the urban slum population in Kochi city significantly. Kochi Urban Poverty Reduction (1998–2004) and Kerala Development Program (1998–2003) reduced the percentage of the slum population from 11% (1996) to 1.37% (2001) in Kochi city.

214 Language and Cross-Cultural Communication in Travel and Tourism

TABLE 11.4 Status of Urban Services in Kochi City.

Indicator	City (municipal corporation)	State (urban)	India (urban)
Percentage of of households with access to tap water (from treated source) within premises	95.55	30.35	84.14
Percentage of households with access to electricity	99.08	97.01	92.68
Percentage of households having toilet facilities within premises	94.62	75.29	72.57
Type of sewerage system*	Underground sewerage system	–	–
Type of solid waste system*	Door to door	–	–
Ownership pattern of housing (%)			
Owned	75.48	88.30	69.16
Rented	21.72	10.00	27.55
Percentage of households living in congested houses	6.05	6.23	32.94

Source: Census of India (2011) Tables of Houses, Household Amenities and Assets, Census of India, 2011.
*District census Handbook, Census of India, 2011.

11.3.9.4 ENVIRONMENT AND LAND USE

Environment and natural resources face the worst horrors of urban development and Kochi is no different; changes in the land use, natural drainage pattern with haphazard development, and the establishment of industries along the water bodies have enkindled the environmental degradation of Kochi, its surrounding areas, and backwaters. The irregular management of solid waste, and lack of proper drainage system can be convicted for polluting the rivers and lakes; water bodies are becoming waste dumping sites, and flowing off the sewage effluents in Canals is a major concern; it pollutes the freshwater and causing eutrophication in the water bodies.

The land is one of the precious natural resources; the increasing population and rapid urbanization are aggravating the demand for land. Changes in land use affect the environment drastically; during the past years, Kochi has seen land-use changes with Urban Sprawl having the largest contribution in engulfing the agricultural lands, paddy fields, and catchment areas of canals. The area under water bodies, low-lying areas/paddy fields, and mixed vegetation is declining whereas the Built area has increased rapidly from 18.20 km^2 in 1967 to 63.20 km^2 (Table 11.5). Kochi

Municipal Corporation has only 0.68% (0.65 km^2) of area devoted to open spaces with 1.07 m^2 per head of open area available to the denizens of the city which is too less when compared to URDPFI and WHO guidelines (Town and Country Planning Organisation, 2015; Karunakaran, 2019).

Coastal areas have always been environmental sensitive; being located close to the sea any activity influencing the environment of the city or its management affects the sea and water body also. Kochi City region is blessed with backwaters, mangrove areas, low lying paddy fields, and canal systems, which became a reason for Kochi's fame are now declared as environmentally sensitive areas due to mismanagement and lackadaisical view of authorities toward the environment. Backwaters and canals of Kochi, which provided the easiest way for transportation, now became sewage dumping sites, which once helped in controlling floods and are now clogged with plastic wastes becoming the reasons for flood and water logging in Kochi. Kochi Estuary receives 57,000 m^3/day of effluent discharge from industries whereas Kochi City disposes 255 million L/day of wastewater in lakes and other water bodies (Department of Town and Country Planning, n.d.; Suchitwa Mission, n.d.).

FIGURE 11.14 Flowing of sewage water from houses directly into canal.

Being one of the largest industrial hubs in the state of Kerala, Kochi is facing a high level of air pollution, and the absence of scientific and environmentally friendly disposal of solid waste is aggravating the problem (Numbudiri, n.d.). The major part of solid and liquid waste finds its way to Vembanad Lake, which also happens to be a Ramasar Site; the fear of loss of wetland ecological system is hovering and pollution is killing its aquatic fauna. The GKA was positioned 24th in the most polluted industrial zone. Kochi is highly susceptible to the changes occurring due to global warming (Kumar et al., 2015); the rise in sea level is engulfing the coastal land (Kerala Perspective Plan 2030, n.d.; K.S.P. Board, 2011); and it is also affecting the livelihood of fishermen; the community lives at the shores and needs a place for better habitation around the coast. Mangalvanam Bird Sanctuary often attributed as the "Green Lungs of Kochi," an ecologically sensitive area lined by thick Mangroves situated amid the city is also facing threats due to urban development (Aziz and Bhupathy, 2006; Preehtha and Oommen, 2016). There is an ardent need to address the environmental changes, a major decision should be made to preserve the environment; development should be done in such a way that it will cause the least harm to the environment; sustainable development approaches should be incorporated.

TABLE 11.5   Land Use Changes in Kochi Municipal Corporation Area.

Land use	1967		1988		2005	
	Area (km²)	%	Area (km²)	%	Area (km²)	%
Water bodies	24.96	26.31	22.07	22.78	20.88	22.01
Low lying areas/paddy fields	9.81	10.34	8.20	8.47	3.01	3.17
Mixed vegetation	41.25	43.47	12.77	13.18	7.14	7.52
Parks/open grounds	0.36	0.38	0.36	0.37	0.65	0.69
Built-up area	18.20	19.18	53.20	54.91	63.20	66.61
Beach	0.30	0.32	0.28	0.29	0	00
	94.88	100	94.88	100	94.88	100

Source: Department of Town and Country Planning (2010) (Draft Development Plan for Kochi, 2031).

11.4   MAJOR CONCERNS

11.4.1   TRANSPORTATION AND TRAFFIC-RELATED ISSUES

Kochi City has a complex network of roads that provides efficient transportation connectivity to both inter-city and intra-city for better transits.

But the insufficient capacity of these networks to carry the passengers in the city affects the efficiency of the road network in and around the city. Reduction in the average speed of vehicles is due to huge traffic created by the high volume of vehicles, lack of efficient management, and inefficiency; lack of transport infrastructure to carry the traffic volume creates congestion, thereby reducing the vehicular speed. The average journey speed of vehicles in major corridors of Kochi city is only 20 km/h (Department of Town and Country Planning, 2010). The lack of pedestrian infrastructure in the CBD area is leading to an increased number of two-wheelers and cars entering the area, which further aggravates the traffic problems. Major roads of the city are overstressed specifically roads connecting the north–south corridor and East–West corridor (City Mobility Plan-Kochi, 2007). The projected data suggests that around 30% of modal switches from road-based transport to the metro will be there on the mainland (Paul et al., 2018). The integration of all different modes of transportation will help in improving the transportation and connectivity of the city especially by improving the water transport within the city by making use of the naturally existing water channels which can provide like Venice and Amsterdam. The integration of different modes will help in improving tourism, economy, and sustainability of the city.

11.4.2 THE DEGRADATION OF URBAN ENVIRONMENT AND POLLUTION

The rising population and change in land use are affecting the environment of Kochi city. The increasing number of vehicles and industries leads to the degradation of air quality. Greater Kochi industrial cluster is one of the highly polluted areas; as per the Comprehensive Environmental Pollution Index (CEPI), Kochi is at the 24th position with a score of 75.08 (Kochi Among Cities with Heavy Pollution, 2012). The urban environment is degrading and industrial suburbs in the city is becoming the hotspot of environmental pollution (Department of Town and Country Planning, 2010). The anthropogenic activities in the city are adversely affecting the marine environment and also resulting in floods, erosion, siltation, etc., which disrupts the environmental balance. The constantly degrading water and air quality is a serious concern; high levels of pollutants in air and water is causing health hazards to the denizens.

11.4.3   *SEWERAGE AND SOLID WASTE MANAGEMENT*

Kochi city lacks efficient sewerage and solid waste management network, which results in the poor performance of waste collection, management, and disposal. The underground sewerage network is absent for the majority of the city area, which can potentially damage the underground water sources. The wastewater collection, treatment, and reuse method are absent, which is discharged directly with limited treatment resulting in the pollution of scarce water resources in the city. Urban Areas of Kochi discharge 3.5 billion litre of sewage into water bodies out of which 90% of this sewage enters the water bodies untreated (Ullas, 2012). Households in Kochi are already facing scarcity of water; further, the groundwater table is also depleting and it is under stress (Suchitra, 2015). The improper management of sewerage and solid waste is becoming a reason for polluting the available resources, if considerable development in the field is not attained it might turn Kochi into a dump yard of human as well as industrial wastes, which will damage the natural resources in the city.

11.5   CONCLUSION

Kochi is a vibrant port city having the potential to become a world-renowned global city. The government initiatives and schemes have helped the city in evolving; ambitious projects like Smart City and Global City concept will improve the urban structure of the city. But, the city also has overwhelming problems to deal with like rising pollution levels, degrading environment, loss of canals, lack of sewage service, slums, oversaturated roads, traffic snarls, etc., which need to be addressed efficiently.

Being a historical city; the old buildings and streets of the city provides a window to sneak into its past; it attracts a considerable tourist population every year. Heritage areas and Tourist Zones should be improved by providing better infrastructure facilities. Conservation of Heritage zones and Buildings is needed; major tourist spots and zones should be connected by making a tourist travel network so that the city can utilize its tourist potential for its economic development. But it should always be kept in mind that contemporary development projects should not cause harm or change to its historical areas and buildings and environment (like backwaters and islands).

The future of Kochi will be highly influenced by its current urban development. Kochi has the potential to become a world-class city; ambitious projects like "Smart city mission" and "Global City" provides a bridge to achieve that target by overcoming the social, economic, and environmental challenges.

KEYWORDS

- **Port City**
- **coastal tourism**
- **culture**
- **heritage**
- **Kerala**

REFERENCES

Action Plan for Greater, Kochi Area. Kerala State Pollution Control Board Action Plan for Greater Kochi Area Executive Summary, 2001.

Anand, A.; Sreevastan, A.; Taraporevala, P. An Overview of the Smart Cities Mission in India, 2018. https://smartnet.niua.org/sites/default/files/resources/scm_policy_brief_28th_aug.pdf

Aziz, P. A.; Bhupathy, S. The Mangalvanam Bird Sanctuary/Mangrove Area, 2006. https://www.researchgate.net/publication/276853762_THE_MANGALVANAM_BIRD_SANCTUARY_MANGROVE_AREA_ERNAKULAM_KERALA_A_brief_report

Census of India, District Census Handbook-Ernakulam, 2011. https://censusindia.gov.in/nada/index.php/catalog/645

City Mobility Plan-Kochi. City Mobility Plan-Kochi, 2007. https://fdocuments.in/document/city-mobility-plan-kochi-2007.html

Delhi Metro Rail Corporation Ltd. Kochi Metro Project: Alwaye-Petta Corridor, 2011. https://kochimetro.org/kmrl_content/uploads/dpr.pdf

Department of Tourism. Kerala Tourism Statistics 2015, 2016. https://www.keralatourism.org/touriststatistics/

Department of Tourism. Kerala Toursim Statistics, 2017. https://www.keralatourism.org/tourismstatistics/tourist_statistics_201720180314122614.pdf

Department of Town and Country Planning. City Developement Plan 2006, n.d. https://cochinmunicipalcorporation.kerala.gov.in/documents/10157/73076ec1-6197-435a-8973-4dd5626e0225

220 Language and Cross-Cultural Communication in Travel and Tourism

Department of Town and Country Planning. District Urbanisation Report 2011, 2011. https://townplanning.kerala.gov.in/town/wp-content/uploads/2019/04/dur_ernakulam.pdf

Department of Town and Country Planning. Draft Development Plan for Kochi City Region 2031, 2010. https://cochinmunicipalcorporation.kerala.gov.in/documents/10157/17825/Vol1_Study%26Analysis.pdf?version=1.0

Doctor of Philosophy. Spatio-Temporal Changes in the Wetland Ecosystem of Cochin City using Remote Sensing and GIS, 2012. https://dyuthi.cusat.ac.in/xmlui/bitstream/handle/purl/3750/Dyuthi-T1712.pdf?sequence=1

Express News Service. Kochi Ranked Topmost Emerging City in Country, 2018. https://www.newindianexpress.com/cities/kochi/2018/aug/02/kochi-ranked-topmost-emerging-city-in-country-1851933.html

G. and G. of India City Sanitation Plan for Kochi. http://www.mohua.gov.in/upload/uploadfiles/files/CSP_Brochure_Kochi.pdf

George, J. An Assessment of Inclusiveness in the Urban Agglomeration of Kochi City: The Need for a Change in Approach of Urban Planning, 2016. https://mpra.ub.uni-muenchen.de/90149/1/MPRA_paper_90149.pdf

Goswami, B.; Kumar, N. A.; George, K. K. Patterns of Commuting for Work: A Case Study of Kochi City, n.d. http://14.139.171.199:8080/xmlui/handle/123456789/600

Government of Kerala. Department of Tourism, Kerala Tourism, 2019. https://www.keralatourism.org/kochi/industrial-capital-kochi.php (accessed 1 Sept 2019)

Government of Kerala Local Self Government Department. Kerala Sustainable Urban Development Project, Volume 2: City Report, **2005**. (https://www.adb.org/sites/default/files/project-document/75496/32300-02-kochi-ind-tacr.pdf)

Government of India. Ministry of Urban Development, The Smart City Challenge: Stage 2, n.d.

Hridya, R. R. et al. Solid Waste Management in Cochin, India: Practices, Challenges and Solutions. *J. Basic Appl. Eng. Res.* **2016**. http://www.mohua.gov.in/upload/uploadfiles/files/CSP_Brochure_Kochi.pdf

Jain, A. K. *The Kochi Muziris Biennale: Its Impact on the Socio-Cultural Aspects of India*; University College London, 2014. https://www.academia.edu/11825972/The_Impact_of_the_Kochi_Muziris_Biennale

Joseph, Y. A Study on Inland Water Transportation in Kochi City Region, 2012. https://www.researchgate.net/publication/281631681_A_Study_on_Inland_Water_Transportation_in_Kochi_City_Region

K.S.P. Board. Economic Review 2010, 2011. https://www.im4change.org/docs/Economic%20Survey%20of%20Karnataka%202010-11.pdf

Karunakaran, B. Kochi Is Growing at Fast Pace But with Less Than 0.5% of Allocated Open Space. *Times of India* **2019**. https://timesofindia.indiatimes.com/city/kochi/no-space-to-breathe-in-kochi/articleshow/69617387.cms

Kerala Perspective Plan 2030, n.d. https://www.ncaer.org/image/userfiles/file/Kerala%202030/KPP-2030-Vol-2.pdf

Kerala Tourism, 2019. https://www.keralatourism.holiday/tourist-attractions/cochin.php (accessed 25 Aug 2019)

Kochi Among Cities with Heavy Pollution. *The Hindu* **2012**. https://www.thehindu.com/news/cities/Kochi/kochi-among-cities-with-heavy-pollution/article3798444.ece

The Saga of Kochi: Cultural and Heritage Tourism Overview 221

Kochi Municipal Corporation. City Development Plan, 2006.

Kumar, A. et al. Emvironmental Issues of Keral a in General and Kochi/Ernakulam in Particular. *J. Environ. Protect.* **2015**.

Local Self Government Department, Kerala Sustainable Urban Development Project, 2005. https://www.adb.org/sites/default/files/project-document/75505/32300-07-ind-tacr.pdf

Ministry of Urban Development. Mission Statements and Guidelines, 2015.

Numbudiri, S. Industrial Zones Under Scanner for Pollution. *Times of India* **n.d.** https://timesofindia.indiatimes.com/city/kochi/industrial-zones-under-scanner-for-pollution/articleshow/61810851.cms

Padmanabhan, N. *Formation of Kerala Society and Culture*; University of Calicut, 2011. http://www.universityofcalicut.info/SDE/VIsem_formation_of_kerala_society_and_culture.pdf

Paul, S.; Aziz, Z.; Ray, I. The Role of Waterways in Promoting Urban Resilience: The Case of Kochi City, 2018. https://icrier.org/pdf/Working_Paper_359.pdf

Population Division. United Nations, Department of Economic and Social Affairs, The World Cities in 2018: Data Booklet, 2018. https://www.un.org/en/development/desa/population/publications/pdf/urbanization/the_worlds_cities_in_2018_data_booklet.pdf

Prakash, B. A. Urban Unemployment: A Study of Kochi City, 2001. http://www.cds.ac.in/krpcds/report/prakash.pdf

Preehtha, N.; Oommen, V. Urbanization and Biodiversity Conservàtion-Issues, Challenges and Potential. In: *Kerala Environment Congress, Centre for Environment and Development*; 2016. http://cedindia.org/wp-content/uploads/2013/08/KEC-2016-Proceedings-2016-full-pages-with-cover.pdf

Sankhe, S.; Vittal, I.; Dobbs, R.; Mohan, A.; Gulati, A.; Ablett, J.; Gupta, S.; Kim, A.; Paul, S.; Sanghvi, A. India's Urban Awakening: Building Inclusive Cities, Sustaining Economic Growth, 2010. http://mohua.gov.in/upload/uploadfiles/files/MGI_india_urbanization_fullreport011.pdf

Sir Robert Bristow. Cochin Saga, 1959. https://books.google.co.in/books?id=jQcdAAAAMAAJ

Suchitra, M. Water Quality Deteriorating in Kochi, Reveals Audit, Down to Earth, 2015. https://www.downtoearth.org.in/news/water/water-quality-deteriorating-in-kochi-reveals-audit-51031

Suchitwa Mission. Environmental and Social Assessment Report, n.d.

Tourism Cochin, 2019. http://www.tourismcochin.com/general_info.html (accessed 14 Aug 2019)

Town and Country Planning Organisation. Urban and Regional Development Plans Formulation and Implementation (URDPFI) Guidelines, 2015. http://tcpo.gov.in/urban-and-regional-development-plans-formulation-and-implementation-urdpfi-guidelines

Trias, B. A.; Gommans, J. J. L. *Living at the Gates of History*; Leiden University, n.d. https://www.academia.edu/4394447/Living_at_the_Gates_of_History

Ullas, M. A. Population Growth and Environmental Impacts in Kerala. *IOSR J. Humanit. Soc. Sci.* **2012**, *6*, 34–38. https://www.iosrjournals.org/iosr-jhss/papers/Vol6-issue1/E0613438.pdf

CHAPTER 12

Rural Tourism in India: Constraints and Opportunities

C. MAGESH KUMAR, K. SUJATHA, and K. RAJESH KUMAR

Department of Business Administration, Annamalai University, Chidambaram, Tamil Nadu, India

ABSTRACT

Rural areas constitute an important segment of the overall economy of a country. Rural tourism plays a vital role in the development of rural areas and their socioeconomic conditions as well. India is a multidestination country with a wide variety of tourism resources. Rural tourism is a diversified destination as it comprises ecotourism, ethnic tourism, agro tourism, cultural tourism, and historical tourism. In recent years, rural tourism has played a significant role in the growth of the tourism sector in India. This article aims to review several studies and will provide significant views on the constraints and opportunities of rural tourism in India, with some suggestions for policymakers to improve their policies to develop rural tourism.

12.1 INTRODUCTION

In the present globalized era, tourism has acquired a significant position in the economic growth of a country. The tourism sector has emerged as a growth engine for the economic growth of India (Mishra et al., 2020; Rout et al., 2016). The development of this tourism sector will bring economic growth as well as generate employment, foreign exchange earnings,

---

Language and Cross-Cultural Communication in Travel and Tourism: Strategic Adaptations.
Soumya Sankar Ghosh, Debanjali Roy, Tanmoy Putatunda, & Nilanjan Ray (Eds.)
© 2025 Apple Academic Press, Inc. Co-published with CRC Press (Taylor & Francis)

preserve the national heritage and environment, and promote peace and stability (Kamashetty and Gadad, n.d.). In the case of Indian tourism, the instability was created by the seasonal effect, which plays a vital role in lowering foreign tourist arrivals during the off-season periods (Kumar and Singh, 2019). To counter this seasonal effect in India, rural tourism is an essential mode to ensure tourist visits round the year as well as repeat visits (Mishra et al., 2018).

Also, in the past few decades, there has been a rise in European countries' tourist patterns, shifting from mass tourism to alternative forms of seeking relaxation in nature and rural areas (Ana, 2017). In the UK (Middleton, 1982), a shifting pattern of tourists takes place to visit the countryside and rural areas that have the greatest attraction for tourists. Rural tourism has the potential to improve the socioeconomic development of the local community and also preserve the regional historic sites and the local culture in rural areas (Kataya, 2021; Leonandri and Rosmadi, 2018; Slusariuc, 2018). So, it is essential to look into rural tourism in India for its development. Hence, this study will present significant views on the constraints and opportunities of rural tourism based on previous studies in India.

12.2 RURAL TOURISM IN INDIA

India has a wide variety of tourism resources and is often called a "multi-destination country". The rural areas of India have diversified nature, climate, wildlife, landscapes, religious, art, culture, heritage, historical, and archaeological resources. Approximately 7-lakh villages are spread across India, and almost 74% of the population resides in these rural areas. Rural tourism has the potential to improve the socioeconomic condition of rural areas as well as the rural community's lifestyle. The urbancentric approach, which makes the urban lifestyle stressful, leads to an interest in the rural areas and also becomes a counter-urbanization syndrome (Nagaraju and Chandrashekara, 2014).

Since the past few decades, there has been an increase in tourists visiting the art, cultural, heritage, historical, and archaeological sites that are the rural assets of India, making the paradigm shift towards a sustainable tourism approach (Mishra, 2017). Rural tourism has become an economic asset for India as well as a vital tool for poverty alleviation, employment generation, and remote area development (Indolia, 2013; Ray et al., 2012; Verma, 2017). To support this rural tourism development, rural communities and stakeholders are participating in the competitive strategies of rural destinations (Yavana Rani, n.d.). Rural tourism comprises ecotourism, ethnic tourism, agro tourism, cultural tourism, and historical tourism as its dimensions.

12.3 DIMENSIONS OF RURAL TOURISM IN INDIA

FIGURE 12.1 Dimensions of rural tourism.
Source: The authors created it based on the reviews.

12.4 ECOTOURISM

India is known as a megadiversity country due to its diverse ecosystems and unique flora and fauna. The Northeast Himalayas and the Western Ghats are the two most important biodiversity zones in the country. Across

India, there are also many national parks, wildlife sanctuaries, conservation reserves, Ramsar wetland sites, biosphere reserves, tiger reserves, elephant reserves, Natural World Heritage Sites, and important bird areas. These abundant natural features pave the way for potential ecotourism in India (Cabral and Dhar, 2019). Ecotourism is nature-based tourism that helps to develop local communities, generate revenue, and sustain the environment in an eco-friendly manner (Deb, 2019; Karmakar, 2011).

12.5 ETHNIC TOURISM

Ethnic tourism deals with the heritage and cultural sites that are related to the indigenous and exotic communities, also known as ethnic minorities. In India, there are numerous ethnic minorities, most of whom live mostly in mountainous or hilly regions. Kolli Hills, Pachaimalai, and Kalrayan Hills are some of the hill destinations in the Eastern Ghats of India. Ethnic tourism helps to promote development in the local community by enhancing their socioeconomic conditions (Naseer and Palanichamy, 2016).

12.6 AGRO TOURISM

India is an agrarian country where more than 60% of the residents depend on agriculture, and numerous industries are present that depend on it. Agro tourism connects urban tourists to nature by allowing them to participate in agricultural operations in the field and rural activities that entertain and educate them about agriculture (Chandrashekhara, 2018; Khangarot and Sahu, 2019; Kumbhar, 2009). Agro tourism provides tourists with something to see, something to do, and something to buy (Gopal et al., 2008). This leads to the contribution of the rural community and increases their income with sustainable livelihood (Joshi et al., 2020).

12.7 CULTURAL TOURISM

Cultural tourism is associated with the lifestyle, art, culture, and heritage of the community in the concerned location. In India, there are distinctive cultures across the country that have the potential to attract tourists from other countries to visit these sites. This enables a cultural exchange among foreign tourists, which helps to pave the way for attaining universal peace and harmony. Cultural tourism enhances the development of the local community and conserves cultural resources (Ahamed, 2017; Shankar, 2015).

12.8 HISTORICAL TOURISM

India has numerous historical sites, monuments, museums, archaeological sites, and excavations that tourists are attracted to visiting. The discovery of the Harappan Civilization, an iconic settlement site, made that rural area eminent due to rural tourism development. Historical tourism has the potential to develop the socioeconomic conditions of the rural community through their participation and involvement in the steps of enhancing the experience of tourists (Walia and Dar, 2020).

Furthermore, this article discusses the constraints and opportunities of rural tourism in India.

12.9 CONSTRAINTS IN INDIAN RURAL TOURISM

Indian rural tourism has major constraints in its growth process. The notable ones among them (Mohanty, 2014; Ramakumar and Shinde, 2008; Singh, 2020) are the lack of trained manpower, insufficient financial support, lack of local involvement, illiterate population, lack of communication skills, lack of infrastructure, language hindrance, lack of basic business planning skills, absence of supporting industry, traditional beliefs, and fear of exploitation of the rural environment. There is also another set of obstacles that prevail in Indian rural tourism, namely lack of technological updates, lack of access to overseas markets, weak labor policy, low purchasing power of the domestic market, and political intervention (Raj et al., 2013).

Certain constraints in agro tourism (Nimase, 2020) are lack of commercial approach among small farmers; ignorance of farmers in these activities; the presence of an unorganized sector; ensuring hygiene for urban visitors; the consistency of drought that prevails in many areas; most of the farmers holding small-sized and low-quality land; and lack of capital to improve the basic infrastructure. The barriers to rural tourism in the mountainous states of India were evidenced (Kala and Bagri, 2018), namely non-involvement of the local community, lack of knowledge, lack of basic education, poor living conditions, busy daily routine, perceiving tourism seasonality, lack of expertise, power disparities.

The North-Eastern Region (NER) of India, which is geographically characterized by major portions being under hills, and the remaining portions being under plateaus and under plains, has high potential but lags behind in the rural tourism contribution in the country due to the lack of accessibility, lack of transportation facilities, inner line permit system, tourists' wrong perception of security threats based on one or two incidents, limited accommodation facilities only available for the whole region, many destinations are still undiscovered, scarcity of labor, and low level of promotions (Choudhury et al., 2018; Kakati, 2019; Sati, 2019).

The historical tourist sites have issues with resource protection, feasible public access, threats, and maintenance requirements (Binoy, 2011). There is a service quality gap between the rural community and the tourists, which becomes a constraint on rural tourism (Kumra, n.d.). The rural tourism projects initiated across the country by the Government of India result in only a certain number of projects being effective, while the remaining numbers of projects are average and unsuccessful. The non-involvement of the rural community is the main reason for the failure of these rural tourism projects (Trivedi, 2020). Also, the spatial analysis of the rural tourism project sites (Kumar et al., 2017) shows that they are distributed in an imbalanced pattern. These are the significant constraints identified in several studies on rural tourism in India.

12.10 OPPORTUNITIES IN INDIAN RURAL TOURISM

Indian rural tourism has tremendous potential to generate opportunities for the growth process of the country. The various opportunities provided by (Mathur, 2013; Rathore, 2017) are counter-urbanization syndrome, heritage interest, environmental consciousness, level of education, transportation facilities, telecommunication facilities, health consciousness, infrastructure growth, employment generation, exchange of revenue, knowledge enhancement, preservation of culture and tradition, conservation of ecology and the environment. There are also (Talekar and Potdar, 2012) several opportunities created by agro tourism, namely additional income generation for farmers, enhancing the lifestyles of farmers, and

Rural Tourism in India: Constraints and Opportunities 229

preserving natural resources. Also, agro tourism has different avenues, which are farm stays, venturing different agricultural and allied farms, cow milking, rural art, and craft, village fairs and festivals, and animal rides to attract and retain tourists (Mandi et al., n.d.).

In the rural areas, there are tourist-attractive rural tourism products, namely, distinctive architecture, music, arts, folk arts, martial arts, handicrafts, traditional dances, village food, and natural scenery, etc., which can be indulged in through agro tourism. This will be useful for the rural community to sustain when the drought season prevails in the rural areas (Badrinath et al., 2016; Kumar et al., 2020). These rural products generate economic prospects and empower an entrepreneurial approach for the rural communities, which becomes their secondary business and enriches their livelihoods (Saravanan and Rao, n.d.). This will aid in rural development and increase the quality of life for the rural community, reducing migration (Mili, 2012). Detailed information about rural tourism and its products can be reached to tourists and tends to encourage them to visit the site by implementing marketing strategies and promoting the rural tourism destination (Ray and Das, 2011).

Since the northeastern region (NER) of India lags behind in rural tourism, it has a huge opportunity to improve rural tourism. Its rich natural splendor and the grandeur of mountains and rivers, along with its traditions, are its greatest strengths in attracting tourists. The inner line permission system needs to be relaxed for tourists, and transportation facilities should be made available to enrich the rural tourism opportunities in this region (Choudhury et al., 2018; Kakati, 2019; Sati, 2019). The widespread usage of interpretation centers at historical, archaeological, and heritage sites will educate visitors and also assist them in connecting with the culture (Binoy, 2011). The innovation of technology in rural tourism will enable the rural community to move closer to the international tourists, which will enhance their lifestyle and enrich their opportunities in their growth process (Kumar and Shekhar, 2020). These are the potential opportunities identified from the multiple studies of Indian rural tourism.

12.11 CONCLUSION AND SUGGESTIONS

This study presents significant insights into the constraints and opportunities in Indian rural tourism and suggests to policymakers that involving

Self-Help Groups (SHGs) with the rural community will enrich the growth of rural tourism (Shrivastava and Heredge, 2003). Similarly, the Village Tourism Committee should be formed with sub-committees and capacity-building training must be provided to the rural community to enhance rural tourism (Kapur, 2016). In addition to this, the crowdfunding platform should be created by the local community and used as an alternative financing source for investing in rural tourism activities to attract tourists and integrate them into the rural tourism development (Temelkov and Gulev, 2019). Also, the formulation and implementation of marketing strategies will promote the rural tourism destination and improve rural tourism (Ray and Das, 2011). In order to develop rural tourism, the theory of evolutionary approach needs to be adapted for analysis (Streimikiene and Bilan, 2015). The 4As, representing Attractions, Accessibility, Amenities, and Ancillaries, are the assets of tourism. The rural community and government must control these assets to ensure the sustainability of rural tourism in the long run (Sugiama, 2019). Therefore, these suggestions might be prioritized during policy development for rural tourism growth. Thus, this article provides significant views on the constraints and opportunities in Indian rural tourism, as well as some suggestions for policymakers to consider as they design policies to boost Indian rural tourism.

KEYWORDS

- **rural tourism**
- **rural areas**
- **India**
- **constraints**
- **opportunities**
- **development**

REFERENCES

Ahamed, M. Cultural Heritage Tourism—An Analysis with Special Reference to West Bengal, India. *Int. J. Recent Trends Bus. Tour.* **2017,** *1* (4), 55–62.

Ana, M.-I. Ecotourism, Agro-Tourism and Rural Tourism in the European Union. *Cactus Tour. J.* **2017,** *15* (2), 6–14.

Badrinath, V.; Agalya, S.; Abirami, G.; Aarthi, M.; Aishwarya, R. Scope for Promoting Agriculture Cum Rural Tourismin Thanjavur District- A Case Study. *Indian J. Sci. Technol.* **2016,** *9* (27), 1–8. https://doi.org/10.17485/ijst/2016/v9i27/97625

Binoy, T. A. Archaeological and Heritage Tourism Interpretation A Study. *SA J. Tour. Herit.* **2011,** *4* (1), 100–105.

Cabral, C.; Dhar, R. L. Ecotourism Research in India: From an Integrative Literature Review to a Future Research Framework. *J. Ecotour.* **2019.** https://doi.org/10.1080/147 24049.2019.1625359

Chandrashekhara, Y. Agro-Tourism and Employment Opportunities in Karnataka : An Economic Analysis. *Epitome: Int. J. Multidisc. Res.* **2018,** *4* (3), 7–11.

Choudhury, K.; Dutta, P.; Patgiri, S. Rural Tourism of North East India: Prospects and Challenges. *IOSR J. Humanit. Soc. Sci.* **2018,** *23* (2), 69–74. https://doi.org/10.9790/0837-2302046974

Deb, R. Ecotourism in Assam: Promises and Pitfalls. *JSSGIW J. Manag.* **2019,** *VI* (II), 41–52.

Gopal, R.; Varma, S.; Gopinathan, R. Rural Tourism Development: Constraints and Possibilities with a special reference to Agri Tourism. *Conf. Tour.n India—Challenges Ahead* **2008,** 512–523.

Indolia, U. A New Mantra for a Rural Development: Rural Tourism. *Int. J. Educ.* **2013,** *2* (July).

Joshi, S.; Sharma, M.; Singh, R. K. Performance Evaluation of Agro-tourism Clusters using AHP–TOPSIS. *J. Oper. Strat. Planning* **2020,** *3* (1), 7–30. https://doi.org/10.1177/2516600x20928646

Kakati, B. K. Rural Tourist Products: An Alternative for Promotion of Tourism in North East India. *J. Tour.* **2019,** *XX* (1), 33–54.

Kala, D.; Bagri, S. C. Barriers to Local Community Participation in Tourism Development: Evidence from Mountainous State Uttarakhand, India. *Tourism* **2018,** *66*(3), 318–333.

Kamashetty, S. B.; Gadad, A. C. *Tourism Development and Its Impact on the Indian Economy,* n.d.

Kapur, S. Rural Tourism in India: Relevance, Prospects and Promotional Strategies. *Int. J. Tour. Travel* **2016,** *9* (1 and 2), 40–49.

Karmakar, M. Ecotourism and Its Impact on the Regional Economy—A Study of North Bengal (India). *TOURISMOS: Int. Multidisc. J. Tour.* **2011,** *6* (1), 251–270.

Kataya, A. The Impact of Rural Tourism on the Development of Regional Communities. *J. Eastern Eur. Res. Bus. Econ.* **2021,** *2021.* https://doi.org/10.5171/2021.652463

Khangarot, G.; Sahu, P. Agro-Tourism: A Dimension of Sustainable Tourism Development in Rajasthan. *JIMS8M: J. Indian Manage. Strat.* **2019,** 21–26. https://doi.org/10.5958/0973-9343.2019.00030.9

Kumar, A.; Singh, G. Seasonal Effect on Tourism in India. *J. Finance Econ.* **2019,** *7* (2), 48–51. https://doi.org/10.12691/jfe-7-2-1

Kumar, G. S.; Rajesh, R.; Kumar, P. Rural Tourism Development and Promotion in Potential Villages of Tamilnadu. *Int. J. Manage.* **2020,** *11* (10), 122–132. https://doi.org/10.34218/IJM.11.10.2020.013

Kumar, N.; Singh, R.; Aggarwal, A. Spatial Analysis of Rural Tourism Sites in Punjab. *Int. J. Appl. Bus. Econ. Res.* **2017,** *15* (21 (2)), 419–429.

Kumar, S.; Shekhar. Technology and Innovation: Changing Concept of Rural Tourism—A Systematic Review. *Open Geosci.* **2020,** *12* (1), 737–752. https://doi.org/10.1515/geo-2020-0183

Kumbhar, V. Agro-Tourism : A Cash Crop for Farmers in Maharashtra (India). *Munich Personal RePEc Arch.* **2009,** 25187.

Kumra, R. Service Quality in Rural Tourism: A Prescriptive Approach. *Conf. Tour. India—Challenges Ahead* **n.d.,** 424–431.

Leonandri, D.; Rosmadi, M. L. N. The Role of Tourism Village to Increase Local Community Income. *Budapest Int. Res. Crit. Inst. (BIRCI—J)* **2018,** *1* (4), 188–193. https://doi.org/10.33258/birci.v1i4.113

Mandi, K.; Azad, A.; Dutta, S.; Hindorya, P. S. (n.d.). Agro Tourism: Exploring New Avenues in Rural India. *Sci. Agric. Allied Sector: Monthly e Newslett.* **n.d.,** *1* (1), 7–13.

Mathur, P. Rural Tourism: An Effective Tool to Strengthen Rural Infrastructure. *Unnati* **2013,** *1* (Jan–Jun), 70–75.

Middleton, V. T. C. Tourism in Rural Areas. *Tour. Manage.* **1982,** *3* (1), 52–58. https://doi.org/10.1016/0261-5177(82)90026-7

Mili, N. Rural Tourism Development: An Overview of Tourism in the Tipam Phakey Village of Naharkatia in Dibrugarh District, Assam (India). *Int. J. Sci. Res. Pub.* **2012,** *2* (12), 710–712.

Mishra, M. Rural Tourism: A Voyage to the Great Repositories of Living Culture. *Odisha Rev.* **2017,** *Dec*, 1–3.

Mishra, P. K.; Rout, H. B.; Kestwal, A. K. Tourism, Foreign Direct Investment and Economic Growth in India. *Afr. J. Hosp. Tour. Leisure* **2020,** *9* (1), 1–7.

Mishra, P.; Rout, H.; Pradhan, B. Seasonality in Tourism and Forecasting Foreign Tourist Arrivals in India. *Iran. J. Manage. Stud.* **2018,** *11* (4), 629–658. https://doi.org/10.22059/ijms.2018.239718.672776

Mohanty, P. P. Rural Tourism in Odisha-A Panacea for Alternative Tourism: A Case Study of Odisha with Secial Reference to Pipli Village in Puri. *Am. Int. J. Res. Humanit. Arts Soc. Sci.* **2014,** *7* (2), 99–105.

Nagaraju, L. G.; Chandrashekara, B. Rural Tourism and Rural Development in India. *Int. J. Interdisc. Multidisc. Stud.* **2014,** *1* (6), 42–48.

Naseer, C. P.; Palanichamy, V. Ethnic Tourism of Eastern Ghats of Tamil Nadu: A Historical Perspective. *Hist. Res. Lett.* **2016,** *38*, 14–20.

Nimase, A. G. Development of Agro-Tourism in Rural Maharashtra: Challenges and Disturbances. *Aayushi Int. Interdisc. Res. J.* **2020,** *VII* (III), 1–6.

Raj, R. V.; Justy, J.; Anoop, C. Rural Tourism in India: Issues and Challenges in Marketing Strategy of Community Tourism. *Int. J. Manage. Soc. Sci. Res.* **2013,** *2* (3), 1–4.

Ramakumar, A.; Shinde, R. Product Development and Management in Rural Tourism. *Conf. Tour. India—Challenges Ahead* **2008,** 443–452.

Rathore, M. S. Rural Tourism in Rajasthan: An Opportunity for Rural Transformation. *Int. J. Emerg. Trends Inf. Knowl. Manage.* **2017,** *1* (2), 20–24.

Ray, N.; Das, D. K. Marketing Practices for Promotion of Rural Tourism: A Study on Kamarpukur, India. *J. Bus. Econ.* **2011,** *2* (5), 382–396.

Ray, N.; Das, D. K.; Sengupta, P. P.; Ghosh, S. Rural Tourism and It's Impact on Socioeconomic Condition: Evidence from West Bengal, India. *Glob. J. Bus. Res.* **2012,** *6* (2), 11–22.

Rout, H. B.; Mishra, P. K.; Pradhan, B. B. Nexus Between Tourism and Economic Growth: Empirical Evidence from Odisha, India. *Int. J. Appl. Bus. Econ. Res.* **2016,** *14* (11), 7491–7513.

Saravanan, A.; Rao, Y. V. Economic Opportunity Through Rural Tourism: An Empirical Study. *SA J. Tour. Herit.* **n.d.,** *6* (2), 92–107.

Sati, V. P. Potential and Forms of Sustainable Village Tourism in Mizoram, Northeast India. *J. Multidisc. Acad. Tour.* **2019,** *4* (1), 49–62. https://doi.org/10.31822/jomat.527278

Shankar, S. Impact of Heritage Tourism in India—A Case Study. *Int. J. Innov. Res. Inf. Secur.* **2015,** *6* (2), 59–61.

Shrivastava, A.; Heredge, M. *Rural Tourism in the Seraj Valley of Himachal Pradesh,* 2003.

Singh, V. Impact of Rural Tourism Development in India. *Mukt Shabd J.* **2020,** *IX* (VI), 3106–3112.

Slusariuc, G. C. Rural Tourism an Opportunity for Development. *J. Tour.* **2018,** *26.*

Streimikiene, D.; Bilan, Y. Review of Rural Tourism Development Theories. *Transf. Bus. Econ.* **2015,** *14* (2 (35)), 21–34.

Sugiama, A. G. The Sustainable Rural Tourism Asset Development Process Based on Natural and Cultural Conservation. *Adv. Soc. Sci. Educ. Humanit. Res.* **2019,** *354,* 249–253. https://doi.org/10.2991/icastss-19.2019.52

Talekar, P. R.; Potdar, M. B. Potential for Development of Agro-Tourism in Kolhapur District of Maharashtra. *Young Res.* **2012,** *1* (Jan), 23–30.

Temelkov, Z.; Gulev, G. Role of Crowdfunding Platforms in Rural Tourism Development. *ScioBrains* **2019,** *56,* 73–79.

Trivedi, S. Potential and Possibilities of Rural Tourism in Darbhanga District, Bihar. *Int. J. Tour. Hotel Manage.* **2020,** *2* (2), 08–16. https://doi.org/10.22271/27069583.2020. v2.i2a.14

Verma, S. An Exploration on the Possibility of Rural Tourism in India. *Rev. Bus. Technol. Res.* **2017,** *14* (2), 36–45.

Walia, N.; Dar, H. Archaeological Connotation and Rural Tourism Development: A Study of Rakhigarhi. *Zeichen J.* **2020,** *6* (12), 378–394.

Yavana Rani, S. *An Empirical Study on the Stakeholders Support on Rural Tourism and Sustainable Growth in Community Business—A Case Study of Karaikudi, Tamilnadu, India*; n.d.; pp 77–82.

CHAPTER 13

The Effect of Social Media Sites Promoting Tourism Industry: An Indian Perspective

VINOD BHATT and AJAY VERMA

School of Applied Sciences and Languages, VIT Bhopal University, Bhopal, India

ABSTRACT

The aim of the study is to examine modern electronic techniques such as social media and their contribution to the tourism industry. Social media interaction adds valuable inputs to the tourism industry which plays a significant role in its promotion. Now, social media sites are full of information related to tourism which creates competition among tourism companies to deliver the greatest deals to individuals regarding tourist attractions. The present study is executed to examine the factors related to social networking sites to promote or motivate the tourism sector. The influence of the social media network is described in terms of detail quality, detail exactness, and detail convenience on promoting tourism in India regarding awareness, attraction, preference, and response. For data analysis, 255 samples were chosen using stratified random sampling statistical tools such as descriptive statistics, correlation, and regression.

13.1 INTRODUCTION

The Internet has influenced communication in the digital era, whether through virtual channels or online social networking. In today's world,

---

Language and Cross-Cultural Communication in Travel and Tourism: Strategic Adaptations.
Soumya Sankar Ghosh, Debanjali Roy, Tanmoy Putatunda, & Nilanjan Ray (Eds.)
© 2025 Apple Academic Press, Inc. Co-published with CRC Press (Taylor & Francis)

information spreads faster through Social Networking Sites than through any other medium. As a result, tourist businesses, destination marketing organizations, and official tourism boards use a variety of social media methods to promote tourism. With the rapid development of technology, particularly computing, the Internet, and mobile telephony, the world has witnessed a communication revolution in the last few decades. As a result, various media platforms, including print, radio, television, and digital, have collided and influenced one another. It is crucial to comprehend how shifts in technology and consumer behavior affect that make travel information easily accessible and distributed. Because of the enormous extent of data that travelers may have access to, the Internet serves as a crucial forum for information exchange between both customers and industry suppliers in the business (e.g., lodging, transit, and attractions), intermediaries (e.g., trip agencies), monitors (e.g., various administrative bodies of the government), along with numerous nonprofit enterprises, including tourist marketing firms (Werthner and Klein, 1999). The globe is growing increasingly interconnected every day. As most nonprofit and creative organizations have adopted digital technologies to communicate with their stakeholders to know the demand and the perception of the audience. Social media and the Internet have fundamentally changed the relationship between organizations and their audiences (Besana and Esposito, 2019).

13.2 LITERATURE REVIEW

It is important to remember that India's travel and tourism sector ranks seventh globally and accounts for about 9.6% of the country's gross domestic product (GDP). Over the next 10 years, it is anticipated to grow by 6.9% annually, making it the fourth-largest economy worldwide. Infrastructure construction, job creation, and skill improvement are estimated to have a multiplier effect on India's socioeconomic development. Appropriate customer engagement in village lodging, fabric tourism, tourism in tribal areas, and nature tourism is the technique of sustainable tourism in India (Heather Carrerio, 2018) because contemporary culture is so interconnected globally; developing and maintaining an online presence is essential for boosting your business and growing your social network (Bhatt et al., 2018). The hotel and tourism industries have increasingly embraced smart solutions to improve their competitiveness and total quality management as a result of the rapid development of information technology (Buhalis, 2020; Buhalis and Leung, 2018; Claver-Cort es et al., 2008; Llach et al., 2016). Social media has been able to connect trillions of individuals from across the world who engage virtually with one another even when they have never met. (Kavoura Androniki, 2014). Currently, social media is playing a more and bigger role as a source of information for travelers, and potential tourists. Additionally, social media has since taken on a significant role in the online tourist industry by facilitating consumer connections (Ministry of Tourism, 2019). The information interchange between online travelers and the service provider is facilitated by a wide range of technological gateways, including Google search, sites for online booking, and webpages of tourism organizations (Xiang et al., 2008). Tourists create virtual communities, exchange information, and get updates on a diverse range of social media platforms. These include a wide range of Technological tools and are available in a variety of shapes. Wiki pages, blog posts, microblogs, social networks, content communities, review websites, and polling sites are a few of the more well-liked ones (Stillman and McGrath, 2008). In general, social media are Internet-based tools used to share user-generated content, such as media impressions made by users, usually based on pertinent involvement, and preserved or communicated through available online sources for other customers who may be influenced by them (Blackshaw, 2006). Social media's introduction has caused a radical shift in forms of communication globally by enabling users to connect, observe, and share information. Social media's development as a new technology has altered how the tourist industry operates, which in turn has had a big impact on the sustainable tourism industry. (Chatterjee and Dsilva, 2021). Social networks, microblogs, and electronic word of mouth (e-WoM) have become a crucial aspect of daily life. It has a huge impact on both the personal and professional spheres and a new age for the global economy is being ushered in by people and corporations using social media (Trusov et al., 2009). The impact of social networking on organizing trips and communicating travel advice has been studied as an emerging subject (Tung et al. 2011). With the successful usage of social media in various aspects, technology has revived and elevated the corporate environment. In the modern period, the consumer society has more freedom to express their ideas, request services, and make suggestions for the two-way transmission of information (Choudhury and

Mohanty, 2018). Social media refers to a variety of online, word-of-mouth platforms, such as blogs; chat rooms and discussion boards sponsored by businesses; consumer-to-consumer email, websites, and forums for product or service reviews; Internet discussion boards; and more (Mangold and Faulds, 2009). Technological advancements and the use of social media in archival services is a relatively recent phenomenon (Bountouri and Giannakopoulos, 2014).

13.3 OBJECTIVE OF THE RESEARCH

Some general questions have been developed to attain the goal of the study. The influence of the social media network is described in terms of detail quality, detail exactness, and detail convenience on promoting tourism in India regarding awareness, attraction, preference, and response. With this broad area, some subquestions were designed:

1. What effects do social networking sites have on detail quality, detail exactness, and detail convenience in creating interest and inspiring tourism in India?
2. What effects do social networking sites have in its variables in terms of detail quality, detail exactness, and detail convenience in creating interest and inspiring tourism in India?
3. What effects do social networking sites have in terms of detail quality, detail exactness, and detail convenience in creating interest and inspiring tourism in India?
4. What effects do social networking sites have in terms of detail quality, detail exactness, and detail convenience in encouraging tourism in India?

13.4 RESEARCH MODEL

In the present research study, the social media network is an independent variable; however, detail quality, detail exactness, and detail convenience were taken as dependent variables. The current era has experienced a substantial change in the Internet utility and the use of other social network platforms as they reflect substantial changes in family relation and social, economic, and political forms worldwide. All including social scientists and researchers believe that social media and the Internet have entered a new

The Effect of Social Media Sites Promoting Tourism Industry 239

phase of engaging societies globally directly or indirect which results in an exchange of information, ideas, culture, belief, and values.

13.5 SOCIAL MEDIA AND TOURISM INDUSTRY

As it is widely accepted, this is the age of communication, and social media plays an important role in ensuring the flow of communication. The continuous development of communication means adding and incorporating various types of social media sites to achieve the goal of a business organization. The use of social media is revolutionizing traditional promotional, and advertising strategies, and the travel and tourism sector is well known for being an information-intensive sector. (Sheldon, 1997; Werthner, and Klein, 1999). Furthermore, the presence of numerous social media applications has bestowed to the transformation of traditional tourism and marketing methods. The effect of the sites of the social media on tourists are very significant. Social media sites provide necessary information to the tourist about the desired place. It helps direct the client as a satisfied customer to the service. Social networking sites not only describe about places for tourism but also advice about the secure channels for purchasing and booking various trips. Social media sites are anticipating explicit details about prices, travel, and services offered by the related industry, and tourism companies and agents clearly benefit from social media and its video coverage.

13.5.1 FACEBOOK

Facebook is the most common and reliable platform for people who are concerned with communication about politics, religion, and international affairs; policies; and tourism destinations. Most of the research shows that Facebook helps other people to connect deeply with an individual's experience about the place he/she visited. The largest social media sites, such as Facebook, for instance, give business marketers access to real-time statistics that assess the efficacy of their adverts in real-time. Companies and organizations can decide how to communicate with the public and what issues are discussed online by utilizing the available information (van Dijck, 2013, p. 206). In addition to other points, it is possible to consider comments and shares as a more trustworthy method of gauging user participation (Chugh, R, Patel, SB, Patel, N and Ruhi, U 2019). A

240   Language and Cross-Cultural Communication in Travel and Tourism

tourist marketer must have a thorough awareness of the gender, age, geography, and interests of the target audience in order to effectively promote a destination on Facebook. Whether it has access to comprehensive visitor surveys or only has website analytics to work with, be sure to make use of the information it has at its disposal to clearly identify its audiences. Additionally, Facebook will be used as a consumer market research tool by locations to collect user insights, assess the level of interaction with content, and crowd-source concepts before commercializing them (Rahman Shahnoor, 2017).

13.5.2 TWITTER

People use Twitter to learn about new things, make suggestions, and share their personal experiences with the rest of the world. This facilitates what makes a place special, as well as the types of people that want to visit. People enjoy viewing images and videos of areas they want to visit. Twitter is regarded as the most popular microblogging platform in the world and is one of the top 10 most visited websites (Antoniadis et al., 2015; Philander et al., 2016). Also, it is considered one of the most significant sources of recent personal and societal knowledge (Tenkanen et al., 2017). According to Kwak et al. (2010), Twitter's popularity had a tremendous impact on hospitality and tourism. (Kim et al., 2013). Twitter is also a good place to look into e-WoM. Subjects in the tourism industry might use Twitter to get qualitative feedback or set competitive benchmarks.

13.5.3 SNAPCHAT

Snapchat is becoming increasingly popular, and not just among teenagers. It is one of the digital platforms that allow people to share personal updates in a short amount of time, such as taking images, sending video clips, adding identity, and so on.

13.6 VARIABLES OF SOCIAL MEDIA NETWORKING

13.6.1 DETAIL QUALITY

Detail quality is defined as the ability to bring together a group of experiences, competence, knowledge, and attitudes that support carrying out the

The Effect of Social Media Sites Promoting Tourism Industry   241

duties of independents at a definite precision level. The individual ability is considered by the following:

1. Efficiencies connected to computer culture.
2. Ability in the use of computer
3. Capability associated with knowledge culture
4. Effectiveness of dealing with worldwide web programmers and assistance.

It is understood that the accessibility of tourism-related information to independents in social media enhances tourism and paves people to explore different tourist locations for treatment, refreshment, and other activities.

13.6.2 DETAIL EXACTNESS

Detail exactness is known to be error-free knowledge that provides quality decisions by avoiding wrong choices, thus saving time and money. Information exactness plays a crucial role in decision making, according to the researcher, and is meagerly dependent on the estimate, experiment, and organization, while the greater advantage is gained by scientific and reasoning methods for effectiveness in the decision. Thus, information exactness gives trust the tourists to be aware of the purchase choice and touristic sites.

13.6.3 EASY TO USE

Easy to use is known for utilizing and implementing decision making easily. The network of social media is defined as convenient as it gives language, techniques, and knowledge on numerous issues.

13.6.4 AWARENESS

Awareness is pointed out as a crisp around a particular emotion while recognizing the feeling is called perception. Consequently, to the awareness,

242 Language and Cross-Cultural Communication in Travel and Tourism

preference, attraction, and response model, awareness stands to enchant customers' decision to make the purchase.

13.6.5 ATTRACTION

Attraction is described as the procedure of socialization in which natural values are combined with the culture of the community for personality development. The priority of a victorious vendor is on the maker's capacity to clear up the work issues to gain profit.

13.6.6 PREFERENCE

Preference is considered as a change in outlook to a matter the human beings are missing and wish to gain; also, it is a perspective to avoid something or a subject. A victorious manufacturer builds the preference of customers regarding the service and its further development.

13.6.7 RESPOND

Respond is defined as the behavior of a person in which actions are built towards a particular behavior or cause. A marketer builds the customer towards a stage of decision making in the purchase by centering the positives of service.

13.7 GENERAL HYPOTHESIS

The hypothesis was formulated based on the questions and problems of the study:

At the point of touristic boost (awareness, attraction, preference, and response), no statistically significant registered ($\alpha = 0.05$) effects of social media networks on detail quality, detail exactness, and detail convenience in terms of awareness, and easy to use at the points of touristic boost (awareness, attraction, preference, and respond). A collection of subhypotheses arose from this hypothesis, as follows:

Hypothesis 1: In terms of awareness in tourism in India, social media networks do not appear to have any statistically significant ($\alpha = 0.05$) effect on detail quality, detail exactness, and detail convenience.

The Effect of Social Media Sites Promoting Tourism Industry 243

Hypothesis 2: In terms of attraction in tourism in India, social media networks do not contribute significantly ($\alpha = 0.05$) effect to detail quality, detail exactness, or detail convenience.

Hypothesis- 3: In terms of preference for tourism in India, social media networks do not have a statistically significant ($\alpha = 0.05$) effect on quality, exactness, and convenience of details.

Hypothesis- 4: In terms of response, social media networks have no statistically significant ($\alpha = 0.05$) effect concerning detail quality, detail exactness, and detail convenience in tourism in India.

13.8 RESEARCH DESIGN

The study population comprises randomly selected people from India who rely on social networking sites for tourism purposes. To achieve the study objectives, a questionnaire was prepared with the relevant questions and distributed randomly. The questionnaire was segregated into two parts; the first part dealt with the demographic details of the respondent and the second covered questions related to the study.

13.9 SAMPLING METHOD

This survey employs a random sampling method. Primary data is the primary source of information for the study. The questionnaire is the source of the primary data. Collection of data is being done through the following: Primary data resources—a questionnaire is used to collect primary data from customers who come to the supermarkets to shop; Secondary data resources—secondary data are information that has already been processed. Data are gathered from journals, ProQuest, Google, and the company website. This research employs both primary and secondary data sources.

13.10 DATA ANALYSIS

Cronbach's alpha is used to check the internal consistency coefficient to confirm the reliability of the study. For social media networks and tourist inspirations, the reliability coefficient totaled 76% which was measured suitable for the study.

244 Language and Cross-Cultural Communication in Travel and Tourism

TABLE 13.1 Cronbach's Alpha is a Major of the Internal Reliability of the Questionnaire's Dimension.

Dimension	α
Detail quality	0.711
Detail exactness	0.672
Detail convenience	0.753
Network of social media	0.698
Awareness	0.732
Attraction	0.656
Preference	0.636
Respond	0.747
Total	0.763

Looking at the values listed in Table 13.1, the data explore that the variables of the study are high stability percentages of 76%, which is greater than 71%, indicating that the questionnaire was stable. The statistical analysis and hypothesis testing results were obtained.

The means and standard deviations of the independent variables are represented in the table below, which covers the networks of social media (detail quality, detail exactness, and detail convenience). We also show the averages and standard deviations of the dependent variable (the Tourism Inspiration aspects include awareness, attraction, preference, and response).

TABLE 13.2 The Curving and Sourcing Values for the Study's Variables, as Well as Their Means and Standard Deviations.

	Mean	Standard deviation	Curving	Sourcing
Detail quality	3.92	0.69	− 0.894	1.37
Detail exactness	4.12	0.79	− 0.168	− 0.453
Detail convenience	4.02	0.81	− 0.697	0.23
Awareness	3.99	0.78	0.74	0.59
Attraction	4.02	0.75	− 0.213	− 0.549
Preference	4.00	0.64	− 0.045	0.51
Respond	4.35	0.80	− 1.10	2.49

The curving coefficients in Table 13.2 were all less than one, indicating that the values were all below three. The shape of the data distribution is referred to as sourcing which ensures whether it is in the middle or

The Effect of Social Media Sites Promoting Tourism Industry 245

not. Sourcing refers to the amount of time it takes to distribute data in a particular format. Curving the value of data distribution decides if it is abnormal or normal, on the contrary, sourcing means what role does the shape of the data circulation which determines whether the data is in the middle or not?

A high percentage of sourcing occurs at the edges, and vice versa.

HYPOTHESES TESTING

Theoretical Model: The use of social media networks to promote tourism in India is hypothesized not to have a statistically significant ($\alpha = 0.05$) impact on detail quality, detail exactness, and detail convenience at the stages of touristic encouragement (awareness, attraction, preference, and response)as shown in Table 13.3.

TABLE 13.3 In this Table, We Conducted a Multiple Regression Used to Describe the Effect of Social Media Networks Related to Touristic Encouragement (Attention, Interest, Desire, and Action) in India.

Correlation R	Coefficient of determination R2	F	Significance
0.487a	0.294	63.412	0.000

Table 13.3 represents that the coefficient of correlation between social media networks and their variables was (0.48) and that the value of statistics (F) was (63.41) at a level of significance of (0.05) or less, indicating that the coefficient of correlation between social media networks and their variables was significant (0.48). Shows how social media networks and associated factors (detail quality, exactness, and convenience) affect the stages of a tourist attraction (awareness, attraction, preference, and response). As a result, the null hypothesis was found to be false, while the alternative hypothesis was found to be true. Indicating that the social media network and its variables had a statistically significant ($\alpha = 0.05$) effect on tourism encouragement phases in India (awareness, attraction, preference, and response).

Table 13.4 makes it obvious that only information efficiency and information accuracy, which have β-coefficient values of 0.159 and 0.512, respectively, and T statistical values of 3.124 and 0.512, respectively, indicate a statistically significant effect on performance (9.923). These

TABLE 13.4 In This Table, Social Media Networks are Related to the Touristic Encouragement Stages in the Indian Tourism Industry as Measured by Their Regression Coefficients.

	Under-standard coefficients		Standard coefficients	t-test	Significance
	1.971	0.212		13.93	0.000
Detail quality	0.234	0.061	0.159	3.124	0.000
Detail exactness	0.294	0.029	0.512	9.923	0.000
Detail convenience	0.069	0.030	0.142	1.935	0.001

suggested a strong and favorable impact on promoting tourism. It should be emphasized that because there is no data, there is no value for the β-coefficient in relation to the constant. A collection of related hypotheses resulted from this: This led to the following set of theories:

Hypothesis-1: In terms of awareness in tourism in India, social media networks do not appear to have any statistically significant ($\alpha = 0.05$) effect on detail quality, detail exactness, and detail convenience.

A multiple regression analysis was used to evaluate this hypothesis, and the results are reported in Table 13.5.

TABLE 13.5 In This Table, We Evaluated the Effects of Social Media Networks on Tourism in India by Using Multiple Regression Analysis.

Correlation R	Coefficient of determination R2	F	Significance
0.801a	0.593	631.231	0.000

Table 13.5 shows that at the attention to tourism stage, represents that the coefficient of correlation between social media networks and their variable was 0.801, and the null hypothesis that social media networks have a consistently significant ($\alpha = 0.05$) impact on tourism attention in India was rejected by the value of F-test for a level of significance (0.05).

Table 13.6 shows that the impact of social media networks on tourism promotion was statistically significant with β-coefficient values of 0.772 and 0.435, respectively, and the value of t-test is 24.963. This indicated a positive and significant impact on tourism awareness.

Hypothesis 2: In terms of attraction in tourism in India, social media networks do not contribute significant ($\alpha = 0.05$) effect to detail quality, detail exactness, or detail convenience.

The Effect of Social Media Sites Promoting Tourism Industry 247

TABLE 13.6 Social Media Networks' Effects on Their Variables are Discussed at the Stage of India's Tourism in India.

Model	Under-standard coefficient		Standard coefficient	t-test	Significance
	Regression coefficient	Standard error	B		
The constant	0.523	0.224		2.734	0.000
Communication networks	0.883	0.043	0.772	24.963	0.001

A multiple regression analysis was used to evaluate this hypothesis, and the findings are reported in Table 13.7.

TABLE 13.7 Based on the Multiple Regression Analysis of Social Media Networks with Their Variables at the Interest Stage of Indian Tourism, We have the Following Results.

Correlation	Determination coefficient	F	Significance
0.721a	0.495	501.234	0.001

Table 13.7 represents the coefficient of correlation between social media networks and their variables at the stage of interest of tourism was 0.721 and the value of F-test was 501.234 with a level of significance 0.05 and less, as a result, the null hypothesis was found to be false, while the alternative hypothesis was found to be true. This states that there is a statistically significant ($\alpha = 0.05$) effect of social media networks on tourism interest in India.

TABLE 13.8 At the Interest Stage of Tourism in India, the Regression Coefficient of Social Media Networks with Their Variables.

The model	Under-standard coefficient		Standard coefficients	t-test	Significance
	Regression coefficient	standard error	β		
The constant	0.023	0.134		0.148	0.896
Networks of Communication	0.895	0.0214	0.721	21.936	0.0001

Table 13.8 represents that, with β-coefficient of 0.721 and a value of t-test, networks of communication had a statistically significant impact on tourism interest (21.936). As a result, there was a favorable and large impact on the interest stage.

248 Language and Cross-Cultural Communication in Travel and Tourism

Hypothesis 3: In terms of preference for tourism in India, social media networks do not have a statistically significant ($\alpha = 0.05$) effect on quality, exactness, and convenience of details.

A multiple regression analysis was used to evaluate this hypothesis, and the outcomes are reported in Table 13.9.

TABLE 13.9 The Analysis of the Multiple Regression Results for the Effect of Social Media Networks on Desired Tourism Stages in India.

Correlation	Determination coefficient	F	Significance
0.812a	0.623	712.258	0.001

Table 13.9 shows that the coefficient of correlation between social media networks and tourism desire was 0.812, whereas the value of the F-test was 712.258 with a level of significance of α =0.05 or less. As a result, the null hypothesis was found to be false, while the alternative hypothesis was found to be true, which states that social media networks have a statistically significant impact ($\alpha = 0.05$) on tourism desire in India.

TABLE 13.10 Analysis of the Regression Coefficient with Social Media Platforms and India's Desired Tourism Stage.

The model	Under-standard coefficients		Standard coefficients	t-test	Significance
	Regression coefficient	Standard error	β		
The constant	−0.492	0.112		− 2.992	0.001
Communication network	2.0120	0.033	0.801	26.124	0.0000

According to Table 13.10, communication networks had a huge positive impact on the desired stage, with a statistical value of 26.124 and a β-coefficient of (0.801) indicate a statistically significant effect.

Hypothesis 4: In terms of response, social media networks have no statistically significant effect ($\alpha = 0.05$) concerning detail quality, detail exactness, and detail convenience in tourism in India. The findings of a multiple regression analysis were performed to evaluate this hypothesis, and they are shown in Table 13.11.

The Effect of Social Media Sites Promoting Tourism Industry 249

TABLE 13.11 This Table Studied the Impact of Social Media Networks on Tourism in India Using Multiple Regression Analysis.

Correlation	Determination coefficient	F	Significance
0.434a	0.325	145.213	0.001

The analysis of Table 13.11 depicts the coefficient of correlation between social media networks and tourism action was 0.434, with an F-statistical value of 145.213 and a level of significance $\alpha =0.05$ or less. As a result, the null hypothesis was found to be false, while the alternative hypothesis was found to be true, which positions that a statistically significant difference does not exist ($\alpha = 0.05$) of social media networks with its variables (detail quality, detail exactness, and detail convenience) to encourage tourism at the acting stage.

TABLE 13.12 In This Table, the Regression Coefficient Determines the Effect of Social Media Networks on Tourism in India.

The model	Under-standard coefficients		Standard coefficients	t-test	Significance
	Regression coefficient	Standard error	β		
The constant	2.456	0.1325		14.901	0.001
Communication network	0.501	0.041	0.423	13.962	0.000

Table 13.12 displays the action stage in communication networks and its positive significance influence, with a β-value of coefficient 0.423 and a value of t-test 13.962, indicating a significant and positive effect.

13.11 CONCLUSION AND RECOMMENDATIONS

At the interest stage in supporting tourism in India, social media networks influenced detail quality, detail exactness, and detail convenience. The terminology used on social media platforms greatly influenced the tourist attraction process. The influence of social media networks and their variables is crucial in India. Networks are important and useful instruments for obtaining quick and detailed information about a tourism area in India.

Following are some suggestions based on the findings of the study: All individuals interested in tourism places should be able to see and access

250 Language and Cross-Cultural Communication in Travel and Tourism

sufficient, correct, and easy-to-use information on social media networks, which will lead to an exact decision to obtain the expected touristic service that satisfies the visitors' desires everywhere. The government agencies, the tourism-related sector, board members, and all touristic travel agencies should develop systematic plans to proficiently use social media networks in terms of detail exactness, detail convenience, and detail quality to increase tourist attraction, awareness, preference, and the decision to adopt the relevant touristic service. Distribute specialized photos and brochures about tourism destinations and post them on social media sites so that tourists can easily see all segments and receive appropriate services, as well as identify the desired benefits from each site.

KEYWORDS

- modern electronics techniques
- social networking sites
- virtual interaction
- tourism industry

REFERENCES

Androniki, K. Social Media, Online Imagined Communities and Communication Research. *Library Rev.* **2014,** *63* (6/7), 490–504. http://dx.doi.org/10.1108/LR-06-2014-0076

Antoniadis, K.; Zafiropoulos, K.; Varna, V. Communities of Followers in Tourism Twitter Accounts of European Countries. *Eur. J. Tour. Hosp. Recreat.* **2015,** *6* (1), 11–26.

Aydin, D.; Ömüriş, E. The Mediating Role of Meaning in Life in the Relationship Between Memorable Tourism Experiences and Subjective Well-Being. *Adv. Hosp. Tour. Res.* **2020,** *9100,* 314–337.

Backshaw, P.; Nazzaro, M. *Consumer-Generated Media (CGM) 101: Word-of-Mouth in the Age of the Web-Fortified Consumer*; Nielsen BuzzMetrics: New York, 2006.

Besana, A.; Esposito, A. Fundraising, Social Media and Tourism in American Symphony Orchestras and Opera Houses. *Bus. Econ.* **2019,** *54* (2), 137–144. https://doi.org/10.1057/s11369-019-00118-7

Bhatt, V.; Goyal, K.; Yadav, A. The Authenticity of Social Media Information Among Youth: Indian Perspective. *J. Content Community Commun.* **2018,** *4* (8), 42–45.

Chatterjee, J.; Dsilva, N. R. A Study on the Role of Social Media in Promoting Sustainable Tourism in the States of Assam and Odisha. *Tour. Crit. Pract. Theory* **2021**, *2* (1), 74–90.

Choudhury, R.; Mohanty, P. Strategic Use of Social Media in Tourism Marketing: A Comparative Analysis of Official Tourism Boards. *Atna—J. Tour. Stud.* **2018**, *13* (2), 41–56.

Curlin, T.; Jaković, B.; Miloloža, I. Twitter Usage in Tourism: Literature Review. *Bus. Syst. Res.* **2019**, *10* (1), 102–119.

Lim, Y.; Chung, Y.; Weaver, P. A. The Impact of Social Media on Destination Branding: Consumer-Generated Videos versus Destination Marketer-Generated Videos. *J. Vacat. Mark.* **2012**, *18* (3), 197–206.

Mangold, W. G.; Faulds, D. J. Social Media: The New Hybrid Element of the Promotion Mix. *Bus. Horiz.* **2009**, *52* (4), 357–365.

Ministry of Tourism. *Social Media as an Influencer Among Foreign Tourists Visiting India. Minist. Tour. Gov. India* **2019**, *1*.

Sheldon, P. J. *Tourism Information Technology*; CAB International: Oxon, 1997.

Sigala, M. Web 2.0, Social Marketing Strategies and Distribution Channels for City Destinations: Enhancing the Participatory Role of Travellers and Exploiting Their Collective Intelligence. In: *Web Technologies: Concepts, Methodologies, Tools, and Applications*; IGI Global, 2010; pp 1249–1273.

Stillman, L.; McGrath, J. Is It Web 2.0 Or Is It Better Information and Knowledge That We Need? *Aust Soc Work* **2008**, *61* (4), 421–428.

Tenkanen, H.; Di Minin, E.; Heikinheimo, V.; Hausmann, A.; Herbst, M.; Kajala, L. et al. Instagram, Flickr, or Twitter: Assessing the Usability of Social Media Data for Visitor Monitoring in Protected Areas. *Sci Rep.* **2017**, *7* (1), 1–11.

Trusov, M.; Bucklin, R. E.; Pauwels, K. Do You Want to Be My "Friend"? Monetary Value of Word-of-Mouth Marketing in Online Communities. *Gfk Mark. Intell. Rev.* **2018**, *2* (1), 26–33.

Trusov, M.; Bucklin, R. E.; Pauwels, K. Effects of Word-of-Mouth versus Traditional Marketing: Findings from an Internet Social Networking Site. *J. Mark.* **2009**, *73* (5), 90–102.

Werthner, H.; Klein, S. *Information Technology and Tourism—A Challenging Relationship*; 1999 (Issue Jan). https://doi.org/10.1007/978-3-7091-6363-4

Xiang, Z.; Wöber, K.; Fesenmaier, D. R. Representation of the Online Tourism Domain in Search Engines. *J. Travel Res.* **2008**, *47* (2), 137–150.

Index

A

Alaska
  Alaska Steamship Company, 156
  cruise industry
    American perceptions, 157
    Humboldt Lines, 156
    Southeast Alaska travel brochures, 158
    White Pass summit, 157
  curio shops, 164
    Alaska Indian Basketry, 167
    clubs, 165
    The Daily Alaskan, 165
    The Jeweler's Circular, 165
  Grand European Tour, 151
  Greek sculpture, 153
  northwest coast, 153
  population statistics, 155
  portable cameras, 154
  White gaze
    British Columbia, 158
    Cassier gold rush, 161
    curio shop bot (SIC), 161
    Equimaux (Eskimo), 164
    good buildings, 163
    Ketchikan, 159
    Norwegian Settlement, 162
    Sheldon Jackson's workshop, 159
    Southeast, 160
    Yukon Territory, 158
Ananda Bazar Patrika, 140
Appraisal framework, 173

B

Barisha Club Durga Puja committee, 86–87
Behala 11 Pally Sarbajanin, 85

C

Case studies
  #Dekhoapnadesh, 62–63
  #MyIndia, 63

#ReasonToTravel campaign, 63–64
#TravelForIndia campaign, 62
Chai, 141
Colonial travel narratives
  re-routing, 133–138
Cruise industry
  American perceptions, 157
  Humboldt Lines, 156
  Southeast Alaska travel brochures, 158
  White Pass summit, 157
Cultural diversity, 34
  philosophy of tourism, 35, 36
    safety requirements, 37
    selfappreciation requirements, 38
    self-realization, 38
    social requirements, 37
  tourism and
    cultural tolerance, 43–44
    delinquent framework, 38–39
    diversity and economic growth, 39–40
    financial presentation, 40–41
    sociocultural diversity, 41–43
Cultural tolerance, 43–44
Curio shop bot (SIC), 161
Curio shops, 164
  Alaska Indian Basketry, 167
  clubs, 165
  The Daily Alaskan, 165
  The Jeweler's Circular, 165

D

Delinquent framework, 38–39
Diversity and economic growth, 39–40
Dream of woman
  story background, 104–105
    and discussion, 112–114
Durga puja festival
  community, 72–75
  Hazra Park Puja, 77–78
  history and emergence, 71–72

Index

North Kolkata's Kashi Bose Lane in
2020, 83
organizers, 79
sacred notion, 75–76
Santosh Mitra Square, 80–81

E

Emile Durkheim, 70
Durga puja festival
community, 72–75
Hazra Park Puja, 77–78
history and emergence, 71–72
North Kolkata's Kashi Bose Lane in
2020, 83
organizers, 79
sacred notion, 75–76
Santosh Mitra Square, 80–81
reconstructing, 83
Barisha Club Durga Puja committee,
86–87
Behala 11 Pally Sarbajanin, 85
Naktala Udayan Sangha, 84
Santoshpur Lake Pally, 84
sacred and profane distinction, 70–71
Equimaux (Eskimo), 164

F

France
awareness campaign posters
materials collected, 19–20
results and discussion, 20–25

G

Gallipoli Campaign, 103, 104
Grand European Tour, 151
Greek sculpture, 153

H

HANDETOUR, 180

I

Intercultural adaptation model, 54
Intercultural communication, 49
case studies
#Dekhoapnadesh, 62–63

#MyIndia, 63
#ReasonToTravel campaign, 63–64
#TravelForIndia campaign, 62
experimental methods and materials
literature review, 52–53
research, 54
FICCI report, 51
regulatory and fiscal policies, 52
research methodology, 55
results and discussion
findings, 55
theoretical framework
communication accommodation
theory, 54
intercultural adaptation model, 54
research questions, 54–55
travel and tourism sector, 50
travel bloggers
content analysis, 60, 61–62
facebook page content analysis,
56–59
thematic analysis, 61

J

Japan
awareness campaign posters
materials collected, 14
results and discussion, 14–19

L

Lesser-known indian spaces
colonial travel narratives
re-routing, 133–138
shift
Ananda Bazar Patrika, 140
Chai, 141
travel perspectives, 141–145
in travel perspectives, 138–142

N

Naktala Udayan Sangha, 84

O

Overtourism, 91
COVID-19 pandemic, 92

Index

255

implications, 98–99
literature review, 93–95
materials and methods, 95
results and discussion, 95–96
    French Environment and Energy Management Agency (ADEME), 97

P

Passenger identity
frameworks
    communication model, 2–3
France, awareness campaign posters
    materials collected, 19–20
    results and discussion, 20–25
Japan, awareness campaign posters
    materials collected, 14
    results and discussion, 14–19
levels and subjects
    pattern 3: W→ (S1→ (S2→ TEXT→ H2) H1)→ R, 12
    pattern 4: W→ (S1→ TEXT→ H1)→ R AND W→ (S1→ (S2→ TEXT→ H2) H1)→ R, 12–14
    pattern 1: W→ (S1→TEXT→H1)R, 10–11
    pattern 2: W→ (S1→TEXT→H1)R, 11–12
methods for
    imaginary world level, 8–10
    poster discourse level, 7–8
    utterances, 6–7
results
    communication models, 25–26
    images used, 26
    poster, contents, 26–27
    targeted acts, 27–28
Yamaoka's (2005) communication
    model, 3–4
    modification, 4–6

R

Rural tourism in India, 223, 224–225
agro, 226
constraints, 227–228
cultural, 226–227
dimensions, 225

ecotourism, 225–226
ethnic, 226
historical, 227
opportunities, 228–229

S

Sacred and profane distinction, 70–71
Saga of Kochi, 191
major concerns
    degradation, 217
    sewerage, 218
    solid waste management, 218
    traffic-related issues, 216–217
    transportation, 216–217
methodology, 193–194
study area, 194
    economic development, 204–205
    geography, 194–195
    heritage, 205–206
    location, 194–195
    planning, 206–208
    smart city planning, 208–209
    tourism, 196–201
    trade induced city's growth, 196–201
    transportation, 209–210
    urban development, 206–208
    urban infrastructure, 210
    urban population, 201–204
urban infrastructure
    environment, 214–216
    land use, 214–216
    sanitation, 211–212
    sewage, 211–212
    slums and housing, 213–214
    water supply, 210–211
Santoshpur Lake Pally, 84
Selfappreciation requirements, 38
Self-realization, 38
Sheldon Jackson's workshop, 159
Slums and housing, 213–214
Smart city planning, 208–209
Social media
tourism industry, effect, 235
    attraction, 242
    awareness, 241–242
    data analysis, 243–245

design, 243
detail quality, 240–241
easy to use, 241
exactness, 241
facebook, 239–240
general hypothesis, 242–243
hypotheses testing, 245–249
literature review, 236–238
objective of the research, 238
preference, 242
research model, 238–239
respond, 242
sampling method, 243
snapchat, 240
twitter, 240
variables, 240
Social requirements, 37
Sociocultural diversity, 41–43
Solid waste management, 218

T

Theoretical framework
  communication accommodation theory, 54
  intercultural adaptation model, 54
  research questions, 54–55
Tourism advertising
  literature review
    advertising discourse, 174–175
    appraisal framework, 176–177
    genre, 175–176
    methodology, 178
    SFLS, 174–175
    visual grammar analysis model, 177–178
  result
    emails, 182
    endorsement/testimonial, 181
    HANDETOUR, 180
    logo and slogan, 180
    multimodal approach, 178–183
    type of tour, 179
  Vietnamese tourism
    move-step structure, 183, 185
Tourism industry
  effect, 235
    attraction, 242

awareness, 241–242
data analysis, 243–245
design, 243
detail quality, 240–241
easy to use, 241
exactness, 241
facebook, 239–240
general hypothesis, 242–243
hypotheses testing, 245–249
literature review, 236–238
objective of the research, 238
preference, 242
research model, 238–239
respond, 242
sampling method, 243
snapchat, 240
twitter, 240
variables, 240
Traffic-related issues, 216–217
Travel bloggers
  content analysis, 60, 61–62
  facebook page content analysis, 56–59
  thematic analysis, 61
Travel writings, 118
  The Hungry Tide, 119
  India
    humanities fieldwork, 119–122
    puri and castillo, 119–122
    travel fictions, 119–122
  travel fictions
    esthetic, 123–125
    Hungry Tide, 125–128
    Jungle Nama, 125–128
    literary esthetic, 128–129
    literary travel, 123–125

U

Urban development, 206–208
Urban infrastructure, 210
  environment, 214–216
  land use, 214–216
  sanitation, 211–212
  sewage, 211–212
  slums and housing, 213–214
  water supply, 210–211
Urban population, 201–204

Index

V

Vietnamese advertising, 174
Vietnamese tourism
  move-step structure, 183, 185
Visual grammar analysis model, 177–178

W

Water supply, 210–211
White gaze
  British Columbia, 158
  Cassier gold rush, 161
  curio shop bot (SIC), 161
  Equimaux (Eskimo), 164
  good buildings, 163
  Ketchikan, 159
  Norwegian Settlement, 162
  Sheldon Jackson's workshop, 159
  Southeast, 160
  Yukon Territory, 158
White Pass summit, 157

Y

Yamaoka's (2005) communication model, 3–4
Yukon Territory, 158